Education Standards and Evolution

Wade S Warren

EDUCATION STANDARDS AND EVOLUTION

Copyright © 2018 Wade S. Warren. All rights reserved. Except for brief quotations in critical publications or reviews, no part of this book may be reproduced in any manner without prior written permission.

Cover and illustrations designed by Michael Williams.
Contributing editors Sarah Caskey and Payton Simmons.

ISBN: 13:978-1986391276, 10:1986391272

FOR CASEY AND CHLOE

TABLE OF CONTENTS

Chapter 1: Introduction to the Louisiana Science Standards   1

Chapter 2: The Form of the Louisiana Science Standards   12

Chapter 3: Middle School Standard 8-MS-LS4-1   18
            The Fossil Record

Chapter 4: Middle School Standard 8-MS-LS4-3   26
            Embryology and Common Ancestry

Chapter 5: High School Standard 8-MS-LS4-1   38
           Common Ancestry, DNA Sequencing,
           Embryology, Vestigial Structures and Fossils

Chapter 6: A Text Case Study: *Campbell Biology*,   60
           11th Edition

CHAPTER ONE

INTRODUCTION TO THE LOUISIANA SCIENCE STANDARDS

THE THREE LOUISIANA SCIENCE STANDARDS evaluated in this book are 8-MS-LS4-1, 8-MS-LS-4-3 (for middle school students), and 8-HS-LS4-1 (for high school students). The performance expectations in these three are exact copies of the nationally recognized "Next Generation Science Standards."[1]

In 2008, the Louisiana Science Education Act (LSEA) was passed by a vote of 94-3 in the House and unanimously in the Senate. Governor Bobby Jindal signed the bill into law on June 25 of that year. The act allows public school teachers to use materials that may not be available in the textbooks to inform students in several areas of controversy, including the teaching of evolutionary biology. The law states that Louisiana public schools shall, "create and foster an environment...that promotes critical thinking skills, logical analysis, and open and objective discussion of scientific theories being studied including, but not limited to, evolution, the origins of life, global warming, and human cloning."

In defense of this legislation, I offered testimony to legislators concerning personal and professional experiences in my own life that reveal the need for such laws. Since the passing of this Louisiana law, many other states have taken steps to try to prevent efforts to suppress the free access to information that does not support the Darwinian model being taught in the public schools. These include Alabama, Minnesota, Mississippi, Missouri, New Mexico, Pennsylvania, South Carolina, Tennessee, and Texas.

I also recently (2016-2017) served on the steering committee assigned to update the Louisiana Science Standards for public school students from kindergarten through 12th grade. My experience on this committee has reminded me of the need for defenders of academic freedom to be involved

---

[1] These are searchable at https://www.nextgenscience.org/standards/standards

in the process.² Although the new standards as a complete package are a huge step forward, the specific standards on Darwinian evolution lack balance. Information in the scientific literature demonstrating weaknesses in the modern evolutionary synthesis has been ignored. In fact, some of this information is already found in textbooks widely used across the nation. This was pointed out to the writers of the standards, but resistance to including unbiased language in the standards was vigorously resisted. ³

As context for the details to follow, here is a portion of a summary letter that I submitted to the media before the final vote on the standards by the Louisiana Board of Elementary and Secondary Education (BESE). Darwin Day (February 12, 2017) happened to coincide in timing, hence the title:

DARWIN DAY PHILOSOPY AND THE IMPACT ON THE PROPOSED NEW LOUISIANA SCIENCE STANDARDS

According to the website darwinday.org, the mission of Darwin Day is to inspire people throughout the globe to reflect and act on the principles of intellectual bravery, perpetual curiosity, scientific thinking, and hunger for truth as embodied in Charles Darwin Darwin's ideas did not fit the scientific dogma of his day, namely that the origin of life and the diversity of species are best explained by the work of a creator. Darwin's willingness to challenge the scientific community of his day with evidence-based thinking has made his name one of the most recognizable in all of the history of science.

The scientific community in the modern era would do well to follow his example by continuing the search for the truth. Dr. Peter Ward, paleontologist and professor at the University of Washington, has suggested

---

² For more details, see
http://www.theadvocate.com/baton_rouge/news/education/article_0cab705c-f223-11e6-9a04-db46afea72aa.html

³ For more details see chapter 6 of this book and see
https://www.evolutionnews.org/2017/03/louisiana-adopts-science-standards-including-louisiana-science-education-act/

that medical treatment, specifically with antibiotics, be denied to those who have publicly expressed doubts about the Darwinian model.[4] This dramatic statement appeared to have been said in a comical fashion but illustrates a larger more disturbing phenomenon among Darwinian proponents today: the attempt to silence Darwinian dissent. Many have raised these concerns. For example, G. Theißen of the Department of Genetics at Friedrich Schiller University, wrote in a 2006 article in *Theory in Biosciences*, 124: 349-369, "It is dangerous to raise attention to the fact that there is no satisfying explanation for macroevolution. One easily becomes a target of orthodox evolutionary biology and a false friend of proponents of non-scientific concepts." Scientists who express skepticism about the power of the Darwinian model to explain the diversity of species are often accused of lacking objectivity because of religious beliefs.

The truth is that the peer-reviewed scientific literature reveals healthy disagreement in the scientific community surrounding Darwinian ideas. An article in *Trends in Ecology and Evolution* from 2008 (23: 121-122) acknowledged that there exists a "healthy debate concerning the sufficiency of neo-Darwinian theory to explain macroevolution." In a paper published in *Nature* in 2014 (514:161-164) titled, "Does evolutionary theory need a rethink," the author says, "The number of biologists calling for change in how evolution is conceptualized is growing rapidly." In fact, despite the possible negative ramifications for doing so, more than 950 scientists have signed, as of November 2016 a statement that says, "We are skeptical of claims for the ability of random mutation and natural selection to account for the complexity of life. Careful examination of the evidence for Darwinian theory should be encouraged." The full list is available online at: http://www.dissentfromdarwin.org/.

In the 11th edition of a popular text, *Campbell Biology* (Pearson Education, 2016), the preface states that "chief among the themes of . . . *Campbell Biology* is evolution. Each chapter. . . includes at least one Evolution section." Given this theme, the authors still manage to acknowledge some of the issues in Darwinian philosophy. For example, in discussing phylogenetics, in chapter 26 (pg. 556), the author writes, "In some cases, however, the morphological divergence between related species can be great and their genetic divergence small." No doubt there are major problems in constructing evolutionary relationships using multiple methods.

---

[4] See https://www.youtube.com/watch?v=Sakmq5L3IiE

In a 2012 (486:460-462) paper in *Nature*, titled "Rewriting Evolution," the author discusses problems with trying to construct Darwin's tree of life using an RNA model. The researchers indicate that their findings "are tearing apart traditional ideas about the animal family tree." Similarly in a 2013 (29:439-441) paper in *Trends in Genetics*, titled "Networks: Expanding Evolutionary Thinking," the authors suggest that "the more we learn about genomes the less tree-like we find their evolutionary history to be." In a 2009 (25:473-474) article in *Trends in Genetics*, the author says that Darwinian tenets such as the tree of life and natural selection as the main driving force of evolution have "crumbled, apparently beyond repair." He goes on to say that "all major tenets of the modern synthesis have been, if not outright overturned, replaced by a new and incomparably more complex vision of the key aspects of evolution."

The process of writing new science standards for students from kindergarten through 12th grade in Louisiana is currently underway. The section on standards for evolution contains no reference to research which calls the Darwinian model into question. I serve on the committee tasked with developing these standards, and I have provided dozens of published peer-reviewed research articles to the writers demonstrating problems with the wording in the standards on evolution. I also made recommendations for wording that would more accurately reflect the current state of the science. For example: in the proposed standards for High School students, the content standard HS-LS4-1 says,

> **"Genetic information provides evidence of evolution. DNA sequences vary among species, but there are many overlaps; in fact, the ongoing branching that produces multiple lines of descent can be inferred by comparing the DNA sequences of different organisms. Such information is also derivable from the similarities and differences in amino acid sequences and from observable anatomical and embryological evidence."**

Even a cursory review of the literature reveals major problems in making these inferences. This information is actually already in most biology texts, as previously noted. I suggested that the following wording be added: "Results may differ depending on the sequences used." The result has been a complete disregard for the information provided and the suggested modifications. The full committee meets again on February 13 in New Orleans at 6026 Paris Avenue, beginning at 9 AM. An explanation for the decisions that have been made will be requested. The public is invited to attend.

In many ways the Darwinian thinkers and writers of the modern era have become the exact opposite of the hopes articulated in Darwin Day's mission statement. Instead, they seem to be trying to create an environment where scientific debate and access to information that disagrees with the Darwinian model is suppressed. Does this exemplify intellectual bravery, perpetual curiosity, scientific thinking, or hunger for truth? I think Charles Darwin would be disappointed in a day celebrating his research and thinking that did not allow for intelligent dissent.

The meeting in New Orleans yielded no positive changes to the language of the standards on evolution, but the story did not end there. One more opportunity was allotted for the public to plead with the members of BESE concerning the language of the standards. BESE met to consider and vote on adopting the full slate of new standards at a public meeting on March 7, 2017. I was not alone at this point in my efforts to ask the board to consider the problems with the standards. Other stakeholders and educators, including one of the members of the steering committee who defected after evidently realizing the harm being done to students, gave testimony to BESE on that day asking for modifications to the standards. Below are excerpts from the report that I made to BESE at this public forum.

PORTIONS OF REPORT TO BESE ON MARCH 7, 2017

I served on the Louisiana Science Standards Committee 2016-1017. My vote was the only dissenting vote concerning the adoption of the proposed standards. The bulk of the document is extremely well-done and will likely facilitate many positive changes. However, the sections on the standards for evolution contain no reference to research that calls the Darwinian model into question. I provided dozens of published peer-reviewed research articles to the writers which demonstrate problems with the wording in the standards on evolution. I also made recommendations for wording that would more accurately reflect the current state of the science. I have provided a copy of a report detailing this information for each of you today.

The result (referencing my work with the writers) was a complete disregard for the information provided and the suggested modifications. At the full committee meeting on February 13 in New Orleans, I requested content-specific explanations for the decisions to ignore the suggested edits. The answers that I received did not address my content questions or the scientific literature that was provided. This is public record and the video can be accessed through the Louisiana Department of Education twitter account.

One of the responses that I received suggested that members of the National Academy of Sciences and other prominent scientists would be uncomfortable with the wording changes that I suggested. This notion may be true. However, many prominent scientists share my concern for the dogmatic presentation of Darwinian philosophy in our schools. I have provided for you a copy of a list of nearly 1,000 names of scientists, including members of the National Academy of Sciences and places like Princeton, MIT, Dartmouth, Stanford, University of California at Berkeley, Emory, Penn State University, and many others who have signed a statement as of 2016 that says, "We are skeptical of claims for the ability of random mutation and natural selection to account for the complexity of life. Careful examination of the evidence for Darwinian theory should be encouraged."

The writers of the science standards did not include any of my suggested edits. I would like to encourage the members of BESE to evaluate the evidence that I have presented and to consider new wording in the standards document somewhere that gives our kids in Louisiana a chance to hear a more accurate presentation of Darwinian philosophy.

TIME FOR THE VOTE

After I had finished my report and several other members of the committee had made their public reports to BESE, it was time for deliberation and the vote. One of the committee members submitted a document to BESE evidently refuting some of my claims. I was not given a copy of this document, and apparently this process bothered one of the members of BESE, Dr. Gary Jones. To his credit, he requested that the member of the steering committee who had submitted this documentation join me at the front of the room to try to flesh out more of the details. After several minutes of conversation in front of the full board, some of the other BESE members began to express concerns over what was happening. At this point, one of the BESE members, Kathy Edmonston, suggested that one way to protect the academic freedom of teachers would be to amend the standards document so that the language of the Louisiana Science Education Act would be included. This amendment was added in a 7-2 vote, and then the entire standards package was approved unanimously.

The specific wording looks like this: B. La. R.S. 17:285.1, known as the 'Science Education Act,' requires the State Board of Elementary and Secondary Education, upon request of a city, parish, or other local public school board, to allow and assist teachers, principals, and other school administrators to create and foster an environment within public elementary and secondary schools that promotes critical thinking skills, logical analysis, and open and objective discussion of scientific theories being studied including, but not limited to, evolution, the origins of life, global warming, and human cloning. Such assistance shall include support and guidance for teachers regarding effective ways to help students understand, analyze, critique, and objectively review scientific theories being studied. A teacher shall teach the Louisiana State Standards using the standard textbook and/or instructional materials supplied by the school system and thereafter may use supplemental textbooks and other instructional materials to help students understand, analyze, critique, and review scientific theories in an objective manner, as permitted by the city, parish, or other local public school board unless otherwise prohibited by the State Board of Elementary and Secondary Education. This law shall not be construed to promote any religious doctrine, promote discrimination for or against a particular set of religious beliefs, or promote discrimination for or against religion or nonreligion.[5]

---

[5] For additional information please see: Louisiana Board of Elementary and Secondary Education, Bulletin 1962, Science Content Standards, Approved 3/8/17. Part CXXIII. Bulletin 1962—Louisiana Science Content Standards, as amended March 2017,
http://www.boarddocs.com/la/bese/Board.nsf/files/AK9M2U592E34/$file/AG II_6.3_B1962_Amended_mar_2017.pdf

Although the specific standards on evolution are not balanced, the wording above may serve to protect the academic freedom of some educators. It is interesting to note that the language of the actual standards calls for students to "investigate, evaluate, and reason scientifically." I wholeheartedly agree. When a teacher presents material on Darwinian evolution, the lesson should be as objective as possible, relying on the empirical evidence from the text and hopefully from the scientific literature. If we are to be fair to the students, the full picture of the Darwinian evolutionary model should be presented, including scientific evidence for and scientific evidence against its content.

A FINAL THOUGHT BEFORE BEGINNING THE STANDARDS

Before beginning the work on the specifics of the standards concerning evolutionary biology, please consider what I think to be perhaps the most fundamental problem with Darwinian modeling. It is the question of the origin of biological information.[6] The following chapters will provide evidence from the scientific literature that suggests problems at nearly every level of Darwinian thought, but the question of the origin of new biological information was not known to Darwin and is not explained by the modern synthesis. Here's a brief explanation that is expanded later in this book in the critique of the *Campbell Biology* text.

We know that biological information is stored in DNA. Transcribing parts of DNA into RNA and then translating that information into proteins is known as the central dogma:

THE CENTRAL DOGMA – DNA TO RNA TO PROTEIN

Proteins are very complicated molecules that are built using amino acids, like methionine, glycine, serine, etc. There are twenty of these that are used to build proteins in humans. The instructions for building proteins are found in three base sequences, or three letter sequences. So, the three letter sequence TAC on the "DNA" is copied as "AUG" on the RNA, which codes for the amino acid methionine. Each three base sequence then

---

[6] For more details, see: Meyer, Stephen. *Signature in the Cell, DNA and the Evidence for Intelligent Design.* New York : HarperCollins Publishers, 2009.

functions as a part of a coded language that ultimately gives instructions for how to build proteins. If one of the letters in the sequence is wrong, it can cause major problems in the proper construction of the protein/amino acid sequence. For example, sickle-cell anemia is caused by a problem in the building of a protein called hemoglobin. A single base (letter) substitution, also sometimes called a point mutation, causes the wrong amino acid to be put in the protein.

Proteins are really important. It is very difficult, next to impossible in fact, to talk about any structure or function in the human body without talking about proteins. For example, we know that the human skeleton (bones) are nearly 30% collagen, which is a protein. The flesh of our bodies (skeletal muscle) and our hearts (cardiac muscle) are by weight mostly actin and myosin, which are proteins. Physiologists must use the language of proteins in describing the how the brain works, how the immune system works, how kidneys make urine, how muscles contract, and on and on we could go.

The genetic code is a language of three letter (bases) codes that contains the instructions on how to build proteins. The three letter sequences in the DNA molecule that ultimately dictate which amino acid gets added to a growing protein are a language that the cells understand, a language critical for life. Therefore, explaining how the letters land in their particular order is critical in understanding both the origin and diversity of life. If there are chemical forces (natural causes) that can explain the order of the letters, then we are much closer to explaining life in general.

Stephen Meyer, in his book, "Signature in the Cell," explains the significance of the order of the DNA base letters in chapter 11, titled "Self-Organization and Biochemical Predestination." On page 243, he says, "There are also hydrogen bonds stretching horizontally across the molecule between nucleotide bases, forming complementary pairs. . . But notice too that there are no chemical bonds between the bases along the longitudinal axis in the center of the helix." He is pointing out that there is no chemical bonding (natural explanation) that dictates in the language that one particular letter follows another along the longitudinal axis of the molecule. Meyer goes on to say, "There are no significant differential affinities between any of the four bases and the binding sites along the sugar-phosphate backbone. Instead, the same type of chemical bond (an N-glycosidic bond) occurs between the base and the backbone regardless of which base attaches. All four bases are acceptable, none is chemically favored." In other words, there is no chemical bonding explanation for the order of letters either in the longitudinal or the horizontal axes of the DNA molecule. The order of the letters in the code is critical and naturalism has no explanation for this order. With this in mind, it makes more sense why a famous scientists like Francis Crick would propose a non-scientific idea like panspermia in attempt to explain the origin of life. This is the idea that life came from somewhere else and was seeded on earth.

In order to build new structures (e.g., body plans) or any new functional system (e.g., reproductive, digestive), an enormous amount of new genetic information is required. Inquiring minds want to know about the origins of all this information. Concepts like mutation, gene duplication, recombination, RNA world hypothesis, pseudogenes, *Hox* genes, junk DNA, and others all fall short as explanations. In the case of junk DNA, evidence now suggests that this very popular explanation is, in fact, false.[7] The simplistic model of life envisioned by the modern synthesis, a Darwinian view, does not get anywhere near explaining the origin of the sequence of bases in a single DNA molecule. Additionally, it does not offer, in my opinion, an adequate explanation for where all this new information comes from in each group of living things. A number of new ideas are surfacing as science tries to deal with this problem. They include evolutionary developmental biology, self-organization, natural genetic engineering, neutral evolution, and neo-Lamarckism. "None of these have gained much ground as they still don't solve fundamental problems with the Darwinian model. They serve more as an illustration that controversy in the scientific community over evolution exists."[8]

---

[7] Wells, Jonathan, *The Myth of Junk DNA*. Seattle, WA: Discovery Institute Press, 2011.

[8] See http://www.discovery.org/scripts/viewDB/filesDB-download.php?command=download&id=12175

Evolutionary Developmental Biology [Evo-Devo]. Proponents include Jeffery H. Schwartz, professor of anthropology at the University of Pittsburgh. He was the first elected president of the World Academy of Art and Science in 2007. For more information on evo-devo, read: Jeffrey H. Schwartz, "Homeobox genes, fossils, and the origin of species," The Anatomical Record 257, no. 1 (February 15, 1999): 15-31, http://onlinelibrary.wiley.com/doi/10.1002/(SICI)1097-

0185(19990215)257:1%3C15::AID-AR5%3E3.0.CO;2- 8/full. Also see Gerd Müller's Royal Society talk [written abstract and audio of talk] available here: "New trends in evolutionary biology: biological, philosophical and social science perspectives," Conference November 7-9, 2016, Royal Society, https://royalsociety.org/science-events-and-lectures/2016/11/evolutionary-biology/

Self-Organization. Proponents include Stuart Kaufmann, Professor of Biological Sciences, Physics, Astronomy, University of Calgary. For more information, on self-organization, read: "The Adjacent Possible: A Talk with Stuart A. Kauffman," Edge, November 9, 2003, https://www.edge.org/conversation/stuart_a_kauffman-the-adjacent-possible.

Natural Genetic Engineering. This theory is championed by James Shapiro, bacterial geneticist, Professor of Microbiology in the Department of Biochemistry and Molecular Biology, University of Chicago. For more on natural genetic engineering, read: James Shapiro, "A 21st century view of evolution: genome system architecture, repetitive DNA, and natural genetic engineering," Gene 345 (2005): 91-100, http://shapiro.bsd.uchicago.edu/Shapiro.2005.Gene.pdf. Also see his Royal Society talk [written abstract and audio of talk] available here: "New trends in evolutionary biology: biological, philosophical and social science perspectives," Conference November 7-9, 2016, Royal Society, https://royalsociety.org/science-events-and-lectures/2016/11/evolutionary-biology/

Neutral Evolution. Michael Lynch, Distinguished Professor of Biology, Indiana University Bloomington, advances this theory. For more information on neutral evolution, read: Michael Lynch, "The origins of eukaryotic gene structure," Molecular Biology and Evolution 23, no. 2 (February 1, 2006): 450- 468, from abstract, https://academic.oup.com/mbe/article/23/2/450/1119102/The-Origins-of-Eukaryotic-Gene-Structure.

Neo-Lamarkism. Proponents include Professor Eva Jablonka of The Cohn Institute for the History and Philosophy of Science and Ideas, Tel-Aviv University, Israel. For more information on neo-Lamarkism and epigenetic inheritance, see her Royal Society talk [written abstract and audio of talk] available here: "New trends in evolutionary biology: biological, philosophical and social science perspectives," Conference November 7-9, 2016, Royal Society, https://royalsociety.org/science-events-and-lectures/2016/11/evolutionary-biology/

CHAPTER TWO

THE FORM OF THE LOUISIANA SCIENCE STANDARDS

THE STANDARDS ADDRESSED IN THE CHAPTERS to follow will be organized in such a way as to make it easy for teachers to quickly access information related to a specific standard. However, before addressing the specifics of the standards, some general information will be useful.

Any presentation of evolutionary biology that is given to kids should clearly define the term evolution first. As different pieces of evidence are presented, the students should be asked to keep these in the appropriate categories. In some cases authors of texts use the term evolution to refer to **common descent**. This is the idea that the living things that we see today are descended from a single common ancestor somewhere in the distant past. In other cases, scientists use the term evolution to refer to **natural selection**. This is the mechanism that really made Darwin famous. He suggested that it has the power to explain the emergence of all living things. The current Darwinian synthesis includes knowledge of genetics and can be summarized in the following equation:

GENETIC MUTAION + NATURAL SELECTION + TIME = DIVERSITY OF SPECIES

Other times, authors seem to use the word evolution to simply refer to the observation that **populations undergo changes**. To say that populations undergo changes is one thing; to extrapolate this notion to mean that all living things appeared because populations can change is saying quite a bit more. I suspect that on many occasions evolutionary presenters have in mind that all three of these conceptions of what evolution means – common descent, natural selection as the mechanism, and/or populations undergo change – are interchangeable. But I think it's important to think about each piece of evidence and how it fits into one or more of these definitions.

Next, teachers should, in my opinion, also open any unit on evolutionary biology by defining Neo-Darwinism and how this synthesis differs from Darwin's original treatise. Neo-Darwinism is sometimes called the Modern Evolutionary Synthesis. As seen in the equation above, it combines Mendelian genetics with Darwin's thinking about natural selection and time. The most basic definition would say that evolution occurs by natural selection acting on random genetic mutations over long periods of time, and that this process explains what we see both in the fossil record and in the world around us.

The problem, of course, is that most of the textbooks and materials available to teachers today are written by scientists who present only the strengths of the modern synthesis. In order for the teaching of a unit on evolutionary biology to be fair and objective, more information is needed. Most school teachers simply do not have the time to make this a reality for their students. So, here's some general information that I hope will help balance the introduction on the modern synthesis.

I am not convinced that the Darwinian model sufficiently explains the origin and diversity of life. I also don't think that the evidence supports the idea that natural selection has the power to have produced the information necessary to account for the observable life forms. I am not alone. Nearly 1,000 individuals in the scientific community have publicly, as of January 2016, signed a statement that says, "We are skeptical of claims for the ability of random mutation and natural selection to account for the complexity of life. Careful examination of the evidence for Darwinian theory should be encouraged."[1]

Consider the following information from a document titled "Teaching Evolution in Louisiana: What Teachers Need to Know."[2] This document was produced by the Discovery Institutes Center for Science and Culture. Here are some interesting facts about the modern synthesis that the science text will not reveal. "In November 2016, the Royal Society held a meeting entitled 'New Trends in Evolutionary Biology: Biological, Philosophical and Social Science Perspectives.' The Royal Society may be the most prominent

---

[1] A Scientific Dissent from Darwinism. [Online] 2016. https://dissentfromdarwin.org/

[2] See http://www.discovery.org/scripts/viewDB/filesDB-download.php?command=download&id=12175

scientific organization on the planet: it was started by Robert Boyle (one of the founders of modern chemistry) and at one point led by Sir Isaac Newton, often regarded as one of the greatest scientists who ever lived. At the 2016 meeting, many top scientists and scholars gathered to discuss the future of the theory of evolution. These included Nancy Cartwright,[3] Denis Noble,[4] John Dupré,[5] Russell Lande,[6] Kevin Laland,[7] and Patrick Bateson.[8] The opening talk by Austrian biologist Gerd Müller laid out the areas that the modern synthesis has not been able to explain. These include: phenotypic complexity (the origin body plans, i.e., the anatomical and structural features of living creatures), development, complex behaviors, and non-gradual forms or modes of transition, where there are abrupt discontinuities in the fossil record between different types.[9]

Michael J. Behe, a professor in the Biological Sciences Department, at Lehigh University says, 'Adaptive evolution can cause a species to gain, lose, or modify a function; therefore, it is of basic interest to determine whether any of these modes dominates the evolutionary process under particular

---

[3] Professor, University of Durham, UK and University of California, San Diego, USA

[4] Professor Emeritus and co-Director of Computational Physiology, Oxford University. President of the International Union of Physiological Sciences.

[5] Professor, Philosophy of Science, University of Exeter, Fellow of the American Association for the Advancement of Science (AAAS)

[6] Professor, Norwegian University of Science and Technology (NTNU), Center for Biodiversity Dynamics, Member of the United States National Academy of Sciences (NAS)

[7] Professor of Behavioural and Evolutionary Biology at the University of St. Andrews, Fellow of the Royal Society of Biology.

[8] Professor Emeritus of Ethology at the University of Cambridge. Past Biological Secretary and Vice-President of the Royal Society of London

[9] "New trends in evolutionary biology: biological, philosophical and social science perspectives," Conference November 7-9, 2016, Royal Society, https://royalsociety.org/science-events-and-lectures/2016/11/evolutionary-biology/

circumstances. Because mutation occurs at the molecular level, it is necessary to examine the molecular changes produced by the underlying mutation in order to assess whether a given adaptation is best considered as a gain, loss, or modification of function. Although that was once impossible, the advance of molecular biology in the past half century has made it feasible. In this paper, I review molecular changes underlying some adaptations, with a particular emphasis on evolutionary experiments with microbes conducted over the past four decades. I show that by far the most common adaptive changes seen in those examples are due to the loss or modification of a pre-existing molecular function, and I discuss the possible reasons for the prominence of such mutations.' [10]

Lynn Margulis (1938-2011), member of the National Academy of Sciences, says, 'I was taught over and over again that the accumulation of random mutations led to evolutionary change—led to new species. I believed it until I looked for evidence.'[11] There is currently a grab-bag of new proposed evolutionary mechanisms. Some of the alternatives to the modern synthesis that scientists have presented include evolutionary developmental biology, self-organization, natural genetic engineering, neutral evolution, and neo-Lamarckism. None of these five proposed mechanisms are fully developed or widely accepted. They serve to illustrate the controversy in the scientific community over evolution."[12]

STRUCTURE OF THE NEW LOUISIANA SCIENCE STANDARDS

All of the science standards are published online at the following address: https://www.louisianabelieves.com/resources/library/academic-standards.

---

[10] Michael J. Behe, *"Experimental Evolution, Loss-of-Function Mutations, and 'The First Rule of Adaptive Evolution,'"* The Quarterly Review of Biology 85, no. 4 (December 2010): 419. Abstract.

[11] "Discover Interview: Lynn Margulis Says She's Not Controversial, She's Right," Discover, June 17, 2011, accessed December 21, 2017, http://discovermagazine.com/2011/apr/16-interview-lynn-margulis-not-controversial-right.

[12] These five mechanisms are listed in the footnotes at the end of chapter 1.

In the opening file titled, "000 Science Standards – Framing Introduction," the logic and organization of the standards are clearly explained. Here, the educator is told who was involved in the process and what the standards represent.

The first section tells us, "The Louisiana Student Standards for Science were created by over eighty content experts and educators with input from parents and teachers from across the state. Educators envisioned what students should know and be able to do to compete in our communities and created standards that would allow students to do so. The Louisiana Student Standards for Science provide appropriate content for all grades or courses, maintain high expectations and create a logical connection of content across and within grades.

The Louisiana Student Standards for Science represent the knowledge and skills needed for students to successfully transition to postsecondary educations and the workplace. The standards call for students to:

- Apply content knowledge
- Investigate, evaluate, and reason scientifically
- Connect ideas across disciplines

The Louisiana Student Standards do not dictate curriculum or teaching methods. Decisions about how to teach these expectations are left to local districts, schools, and teachers."[13]

The document then goes on to explain how each standard is organized by grade and includes the definitions of the other major components each standard.. Here's a brief synopsis. Each standard includes:

"Performance expectations – define what students should be able to do by the end of the year.

Clarification statements – examples or additional explanation to the performance expectation.

Science and engineering practices – practices that scientists and engineers use when investigating world phenomena and designing solutions

---

[13] For more details see https://www.louisianabelieves.com/resources/library/academic-standards

to problems. Asking questions (science) and defining problems (engineering).

Disciplinary core ideas – the most essential ideas (content) in the major disciplines that students will learn.

Crosscutting concepts – common themes that have application across all disciplines of science and allow students to connect learning within and across grade levels or content areas."[12]

The next chapter begins the formal discussion of the standards that deal with evolution Each chapter will list the standard to be discussed, middle school standards in chapters three and four and then a high school standard in chapter five, highlighting the problematic language. Then a discussion will follow highlighting information that will be useful for the teacher in presenting the weaknesses in the Darwinian model. All of the information is cited in footnotes (with internet addresses where possible) to make it easy for the teacher to access the information being quoted from the scientific literature. This compilation is obviously not an exhaustive list but should serve as a resource to supplement unbalanced wording of the standards and the biased presentations found in most available textbooks.

CHAPTER THREE

MIDDLE SCHOOL STANDARD 8-MS-LS4-1 THE FOSSIL RECORD

PERFORMANCE EXPECTATION: ANALYZE AND INTERPRET data for patterns in the fossil record that document the existence, diversity, extinction, and change of life forms throughout the history of life on Earth under the assumption that natural laws operate today as in the past.

Clarification Statement: Emphasis is on finding patterns of changes in the level of complexity of anatomical structures in organisms and the chronological order of fossil appearance in the rock layers.

Science and Engineering Practices: Analyzing and interpreting data: Analyzing data in 6-8 builds on K-5 experiences and progresses to extending quantitative analysis to investigations, distinguishing between correlation and causation, and basic statistical techniques of data and error analysis. Analyze and interpret data to determine similarities and differences in findings.

Disciplinary Core Ideas: EVIDENCE OF COMMON ANCESTRY AND DIVERSITY - The collection of fossils and their placement in chronological order (e.g., through the location of the sedimentary layers in which they are found or through radioactive dating) is known as the fossil record. It documents the existence, diversity, extinction, and change of many life forms throughout the history of life on Earth.

Crosscutting Concepts: CAUSE AND EFFECT
Empirical evidence is required to differentiate between cause and correlation and make claims about specific causes and effects.

There are four standards in the middle school section on "Biological Evolution: Unity and Diversity." This standard is found in the file called

"08 Science Standards – Grade 8,"[1] the standards for middle school students. In the following discussion, the content area of the standard to be addressed will be highlighted in bold italics.

Problem 1: The Standard does not Recognize Abrupt Appearance in the Fossil Record

Performance Expectation: Analyze and interpret data for patterns in the fossil record that document ***the existence, diversity, extinction, and change*** of life forms throughout the history of life on Earth under the assumption that natural laws operate today as in the past.

The language of the performance expectation in 8-MS-LS4-1 was addressed during my work with the standards committee. I asked that two words be added to the performance expectation, "***abrupt appearance.***" The standard would have been worded as follows: "Performance Expectation - Analyze and interpret data for patterns in the fossil record that document the existence, diversity, extinction, ***abrupt appearance***, and change of life forms throughout the history of life on Earth under the assumption that natural laws operate today as in the past."

I pointed out to the committee that all biology texts that discuss the Cambrian fossils teach this concept to students. For example, the *Campbell Biology* text, one of the most popular texts in the country[2], discusses the Cambrian fossils by saying, "Many present-day animal phyla appear suddenly in fossils formed 535-525 million years ago early in the Cambrian period. This phenomenon is referred to as the Cambrian explosion. It goes on to say, "In a relatively short period of time, predators over 1 meter in length emerged that had claws and other features for capturing prey; simultaneously new defensive adaptations such as sharp spines and heavy body armor, appeared in their prey."[3]

---

[1] See https://www.louisianabelieves.com/resources/library/academic-standards

[2] Pearson. *Campbell Biology 11th edition*. Pearson. [Online] 2017. [Cited: December 31, 2017.] https://www.pearson.com/us/higher-education/program/Urry-Campbell-Biology-Plus-Mastering-Biology-with-Pearson-e-Text-Access-Card-Package-11th-Edition/PGM209351.html.

The authors of the *Campbell Biology* text openly promote Darwinian evolution. In the section titled "Our Hallmark Features," in the preface of the text, the authors say, "Chief among the themes of both Vision and Change and *Campbell's Biology* is evolution. Further, each chapter of this text includes at least one Evolution section that explicitly focuses on evolutionary aspects of the chapter material, and each chapter ends with an Evolution Connection Question."[4]

This text does not suggest that the Cambrian fossil evidence is a flaw in the Darwinian model. However, students should have the opportunity to discuss the implications of fossil research that indicates that many (some say most) of the known animal phyla appear in Cambrian fossil beds without evidence of their Darwinian model ancestors. I address more of the details from the *Campbell Biology* text in chapter 6.

The problems that research in the fossil beds have uncovered are not hard to find in the scientific literature. Please find after "Problem 2" that follows, a list of quotations from scientists and journalists with identifying descriptions who have published work on this topic.[5] The references can be found in the footnotes for each. Where possible, internet addresses are in the footnotes to help teachers access the information quickly. Notice the language addressed in bold italics.

Problem 2: The Evidence from the Fossil Record does not Support the Darwinian Model

Performance Expectation: Analyze and interpret data for ***patterns in the fossil record*** that document the existence, diversity, extinction, and change of life forms throughout the history of life on Earth under the assumption that natural laws operate today as in the past.

---

[3] Urry, Lisa A., et al. *Campbell Biology 11th Edition*. New York : Pearson, 2017.

[4] For a fuller discussion of the *Cambell's Biology* text, please see chapter 6

[5] See http://www.discovery.org/scripts/viewDB/filesDB-download.php?command=download&id=12175

Clarification Statement: Emphasis is on finding patterns of changes in the level of complexity of anatomical structures in organisms and the chronological order of *fossil appearance* in the rock layers.

Disciplinary Core Ideas*:* EVIDENCE OF COMMON ANCESTRY AND DIVERSITY - The collection of *fossils* and their placement in chronological order (e.g., through the location of the sedimentary layers in which they are found or through radioactive dating) is known as the *fossil record*. It documents the existence, diversity, extinction, and change of many life forms throughout the history of life on Earth.

Weaknesses in the Darwinian Model - Fossils

1. **Stephen J. Gould** (1941-2002) was a famous evolutionary biologist who taught at Harvard University and was a member of the National Academy of Sciences. A classic 1972 paper by Gould argued that the history of fossil species includes two features inconsistent with Darwinian gradualism: 1) Stasis—that is most species exhibit no directional change during their tenure on earth, and 2) Sudden appearance.[6] A few years later Gould remarked that "The extreme rarity of transitional forms in the fossil record persists as the trade secret of paleontology. The evolutionary trees that adorn our textbooks have data only at the tips and nodes of their branches; the rest is inference, however reasonable, not the evidence of fossils.[7]

---

[6] "*Punctuated Equilibria, an Alternative to Phyletic Gradualism*," Models in Paleobiology (1972): 82-115,
https://books.google.com/books?hl=en&lr=&id=3ULyAgAAQBAJ&oi=fnd&pg=PA239&dq=punctuated+equilibria+an+alternative+to+phyletic+gradualism+&ots=j_9-FjBqn&sig=aDYDor5cC4wlpQvqJX6YeR0OsTA#v=onepage&q=punctuated%20equilibria%20an%20alternative%20to%20phyle tic%20gradualism&f=false

[7] Stephen J. Gould, "Evolution's Erratic Pace," Natural History 86, no. 5 (1977): 12–16,
https://www.researchgate.net/publication/234661562_Evolution%27s_Erratic_Pace.

2. **Ernst Mayr** (1904-2005) was an evolutionary biologist from Harvard University; and a member of the National Academy of Sciences. Considered one of the greatest evolutionary biologists of the 20th century, Mayr noted, "Wherever we look at the living biota ... discontinuities are overwhelmingly frequent... The discontinuities are even more striking in the fossil record. New species usually appear in the fossil record suddenly, not connected with their ancestors by a series of intermediates."[8]

3. **Gareth Nelson** was a former vertebrate zoology curator at the American Museum for Natural History. "The idea that one can go to the fossil record and expect to empirically recover an ancestor-descendant sequence, be it of species, genera, families, or whatever, has been, and continues to be, a pernicious illusion."[9] The phrase "the fossil record sounds impressive and authoritative." As used by some persons it becomes, as intended, intimidating, taking on the aura of esoteric truth as expounded by an elite class of specialists. But what is it, really, this fossil record? Only data in search of interpretation. All claims to the contrary that I know, and I know of several, are so much superstition.[10]

---

[8] Ernst Mayr, *What Evolution Is* (New York: Basic Books, 2001), 189, https://books.google.com/books?id=Dtg3l5QTAPsC&pg=PA189&dq=species+usually+appear+in+the+fossil+record+suddenly, +not+connected+with+their+ancestors+by+a+series+of+intermediates&hl=en&sa=X&ved=0ahUKEwjz_ZXIm5bVAhVW0GM KHTNBBuEQ6AEIMDAC#v=onepage&q=species%20usually%20appear%20in%20the%20fossil%20record%20suddenly%2C%20not%20connected%20with%20their%20ancestors%20by%20a%20series%20of%20intermediates&f=false.

[9] From a presentation by Gareth Nelson in 1969 to the American Museum of Natural History, quoted on page 709 in: David M. Williams and Malte C. Ebach, "The Reform of Palaeontology and the Rise of Biogeography — 25 Years after 'Ontogeny, Phylogeny, Palaeontology and the Biogenetic Law' (Nelson, 1978)," Journal of Biogeography 31, no. 5 (2004): 685–712, http://onlinelibrary.wiley.com/doi/10.1111/j.1365-2699.2004.01063.x/abstract.

[10] Gareth Nelson, *"Ontogeny, Phylogeny, Palaeontology and the Biogenetic Law,"* Systematic Zoology 27 (1978): 324–345,

4. **David M. Raup**, is a former curator of Geology at the Field Museum of Natural History said, "Darwin ... was embarrassed by the fossil record ... Well, we are now about 120 years after Darwin and the knowledge of the fossil record has been greatly expanded. We now have a quarter of a million fossil species but the situation hasn't changed much. The record of evolution is still surprisingly jerky and, ironically, we have even fewer examples of evolutionary transitions than we had in Darwin's time. By this I mean that some of the classic cases of Darwinian change in the fossil record, such as the evolution of the horse in North America, have had to be discarded or modified as a result of more detailed information -- what appeared to be a nice simple progression when relatively few data were available now appear to be much more complex and much less gradualistic. So Darwin's problem has not been alleviated in the last 120 years and we still have a record which does show change but one that can hardly be looked upon as the most reasonable consequence of natural selection."[11]

5. **Henry Gee** is a vertebrate paleontologist and senior editor for biology at Nature. "We call new fossil discoveries missing links as if the chain of ancestry and descent were a real object for our contemplation, and not what it really is: a completely human invention created after the fact, shaped to accord with human prejudices... No fossil is buried with its birth certificate. That, and the scarcity of fossils, means that it is effectively impossible to link fossils into chains of cause and effect in any valid way... To take a line of fossils and claim that they represent a lineage is not a scientific hypothesis that can be tested, but an assertion that carries the same validity as a bedtime story—amusing, perhaps even instructive, but not scientific."[12]

---

https://www.researchgate.net/publication/31252913_Ontogeny_Phylogeny_Paleontology_and_the_Biogenetic_Law.

[11] David M. Raup, *"Conflicts between Darwin and paleontology,"* Field Museum of Natural History Bulletin 50 (1979): 22–29, https://archive.org/details/cbarchive_121465_conflictsbetweendarwinandpaleo19 30.

[12] Henry Gee, *In Search of Deep Time: Going Beyond the Fossil Record to a Revolutionary Understanding of the History of Life* (London and Ithaca, NY; Comstock Publishing

6. **Eugene Koonin** is an evolutionary molecular biologist and a Senior Investigator for the National Center for Biotechnology Information (National Library of Medicine, National Institutes of Health), Editor-in-Chief of Biology Direct, and a member of the National Academy of Sciences. He has said, "Major transitions in biological evolution show the same pattern of sudden emergence of diverse forms at a new level of complexity. The relationships between major groups within an emergent new class of biological entities are hard to decipher and do not seem to fit the tree pattern that, following Darwin's original proposal, remains the dominant description of biological evolution."[13]

7. **Christian Schwabe** is a Professor of Biochemistry and Molecular Biology at Medical University of South Carolina. Schwabe provides a wholesale critique of modern evolutionary theory. He proposes a type of "embryotic evolution," stating, "When a large number of organisms are found all over the earth within a relatively narrow time limit then life had many origins. A large number of different organisms appearing in their final form with all appendages functioning means that all development had occurred before the animals broke ground. That can only mean evolution occurred in a stem cell and that the animal was essentially finished when it appeared in the fossil record. The well-documented inability to discover intermediate forms again points to the stem cell as the evolutionary unit... The literature abounds with stories and glossy pictures of conversion of deer to whale and conversions of fishes to terrestrial animals. While the skeletons depicted are real, the fleshing out and the into—or from— the water migration are indeterminable functions of the Darwinian hypothesis, not of reality. No segment of the record provides evidence of succession in the line of a genotype and there is no evidence for adaptation beyond the

---

Associates, 1999), 32, 113, 116,
https://books.google.de/books?id=TInB03o5uegC&printsec=frontcover&hl=de#v=onepage&q&f=false.

[13] Eugene V. Koonin, *"The Biological Big Bang model for the major transitions in evolution,"* Biology Direct 2, no. 21 (August 20, 2007),
https://www.ncbi.nlm.nih.gov/pmc/articles/PMC1973067/pdf/1745-6150-2-21.pdf.

limits of the species and the variants that already exist... The fossil record of the evolutionary events supports the conclusions that one must draw at the sight of the Cambrian 'edge' with almost derisive clarity. How much better, Nature would ask, can it be shown that animals do not change appreciably during their existence. Jelly fish are seen in Ediacaran formations dated 600 million years, many arthropods are still alive after 500 million years and modern insects are found frequently as amber inclusions dating back over 100 million years? Are we blindfolded by the paradigm?"[14]

---

[14] Christian Schwabe, *"Embryotic evolution: An ancient question, a new answer,"* Cell Cycle 7, no. 11 (June 1, 2008): 1503– 1506
http://www.tandfonline.com/doi/abs/10.4161/cc.7.11.6357.

CHAPTER FOUR

MIDDLE SCHOOL STANDARD 8-MS-LS4-3 EMBRYOLOGY AND COMMON ANCESTRY

PERFORMANCE EXPECTATION: ANALYZE DISPLAYS OF pictorial data to compare patterns of similarities in the embryological development across multiple species to identify relationships not evident in the fully formed anatomy.

Clarification Statement: Emphasis is on inferring general patterns of **relatedness among embryos** of different organisms by comparing the macroscopic appearance of diagrams or pictures.

Science and Engineering Practices: Analyzing and interpreting data: Analyzing data in 6-8 builds on K-5 experiences and progresses to extending quantitative analysis to investigations, distinguishing between correlation and causation, and basic statistical techniques of data and error analysis. Construct, analyze, and/or interpret graphical displays of data and/or large data sets to identify linear and nonlinear relationships

Disciplinary Core Ideas: EVIDENCE OF **COMMON ANCESTRY** AND DIVERSITY. Anatomical similarities and differences between various organisms living today and between them and organisms in the fossil record, enable the reconstruction of evolutionary history and the inference of lines of evolutionary descent. Comparison of the embryological development of different species also reveals similarities that show relationships not evident in the fully-formed anatomy.

Crosscutting Concepts: PATTERNS. Graphs, charts, and images can be used to identify patterns in data.

There are four standards in the middle school section on "Biological Evolution: Unity and Diversity." This standard is found in the file called

"08 Science Standards – Grade 8,"[1] the standards for middle school students. The content area of the standard to be addressed will be highlighted in bold italics.

Problem 1: The Evidence from Embryology does not Support the Darwinian Model

Performance Expectation: Analyze displays of pictorial data to *compare patterns of similarities in the embryological development across multiple species* to identify relationships not evident in the fully formed anatomy.

Clarification Statement: *Emphasis is on inferring general patterns of relatedness among embryos of different organisms* by comparing the macroscopic appearance of diagrams or pictures.

Disciplinary Core Ideas: EVIDENCE OF COMMON ANCESTRY AND DIVERSITY. Anatomical similarities and differences between various organisms living today and between them and organisms in the fossil record, enable the reconstruction of evolutionary history and the inference of lines of evolutionary descent. *Comparison of the embryological development of different species* also reveals similarities that show relationships not evident in the fully-formed anatomy.

Embryology, simply stated, is the study of an growing embryo. The *Campbell Biology* text[2] states that, "At some stage in their embryonic development, all vertebrates have a tail located posterior to the anus (referred to as a post-anal tail), as well as pharyngeal (throat) arches. Descent from a common ancestor can explain such similarities." This argument was a favorite of Darwin's and made famous by Ernst Haeckel in the late 1800s.

---

[1] See https://www.louisianabelieves.com/resources/library/academic-standards

[2] Urry, Lisa A., et al. *Campbell Biology 11th Edition*. New York : Pearson, 2017, p. 477.

As it turns out, much controversy clouds the use of embryological drawings to suggest relatedness between species.³ The real question is how much of this controversy the students should know about. My answer would be to tell them the entire story. It's a great illustration of the bias and corruption that exists in the scientific community.

Here are comments of well-known scientists accompanied by the citations in the footnotes. Once again internet addresses are provided where possible to aid in easy access.⁴

1. **Jonathan Wells**, in an article about Haeckel' embryos, is quoted as saying, "In The Origin of Species Charles Darwin wrote that 'the embryos of mammals, birds, fishes, and reptiles [are] closely similar, but become, when fully developed, widely dissimilar.' He inferred that all vertebrates 'are the modified descendants of some ancient progenitor,' and that 'the embryonic or larval stages show us, more or less completely, the condition of the progenitor of the whole group in its adult state' (Darwin 1859, pp. 338, 345). Darwin's contemporary Ernst Haeckel called this the 'Biogenetic Law,' according to which 'ontogeny recapitulates phylogeny.' To illustrate the law, Haeckel (1891) produced drawings of vertebrate embryos which have been widely used in biology textbooks ever since (Figure 1). "But Haeckel's Biogenetic Law was discredited by embryologists in Darwin's lifetime (Bowler 1989); recent work has shown that Haeckel's drawings misrepresent the embryos they purport to show (Richardson et al. 1997); and Haeckel entirely omitted the earliest stages of development in which the various classes of vertebrates are morphologically very different (Elinson 1987). Biology teachers should be aware that Haeckel's drawings do not fit the facts."⁵

---

³ For a fuller explanation, see chapter 6.

⁴ See http://www.discovery.org/scripts/viewDB/filesDB-download.php?command=download&id=12175

⁵ J. Wells, *"Haeckel's Embryos & Evolution: Setting the Record Straight,"* American Biology Teacher 61, no. 5 (May 1999): 345-349.

2. **Stephen J. Gould** (1941 – 2002) was a Harvard University evolutionary biologist. He said, "We should therefore not be surprised that Haeckel's drawings entered nineteenth-century textbooks. But we do, I think, have the right to be both astonished and ashamed by the century of mindless recycling that has led to the persistence of these drawings in a large number, if not a majority, of modern textbooks!"[6]

3. **Elizabeth Pennisi** has written an article acknowledging that Haeckel's embryo drawings are both fraudulent and continue to be used in textbooks. It states: "generations of biology students may have been misled by a famous set of drawings of embryos published 123 years ago by the German biologist Ernst Haeckel. They show vertebrate embryos of different animals passing through identical stages of development. But the impression they give, that the embryos are exactly alike, is wrong." The article quotes further quotes leading embryologist **Michael Richardson** by stating that "it looks like it's turning out to be one of the most famous fakes in biology."[7]

4. **Alex T. Kalinka, Karolina M. Varga et al**. Both Kalinka and Varga are of the Max Planck Institute of Molecular Cell Biology and Genetics. This article challenges claims of early similarities between embryos. Many textbooks simply show images of two vertebrate embryos and ask the reader to imagine that since they look similar common descent should be considered as a viable concept. These researchers looked at genes from several different species of Drosophila and found that these genes reflected the hourglass model. And some types of genes adhered more closely than others.[8] The hourglass model imagines that both early and late embryonic forms are more diverse than the intermediate stages of development.

---

[6] Stephen Jay Gould, *"Abscheulich! (Atrocious!),"* Natural History (March 2000).

[7] Elizabeth Pennisi, *"Haeckel's Embryos: Fraud Rediscovered,"* Science 277, no. 5331 (September 5, 1997): 1435a.

[8] Alex T Kalinka, Karolina M Varga et al., *"Gene expression divergence recapitulates the developmental hourglass model,"* Nature 468 (December 9, 2010): 811.

Problem 2: Common Ancestry is not Supported by the Fossil Record

Please notice the language of the standard to be discussed in bold italics from the Disciplinary Core Ideas.

Disciplinary Core Ideas: ***EVIDENCE OF COMMON ANCESTRY AND DIVERSITY.*** Anatomical similarities and differences between various organisms living today and between them and organisms in the fossil record, enable the reconstruction of evolutionary history and the inference of lines of evolutionary descent. Comparison of the embryological development of different species also reveals similarities that show relationships not evident in the fully-formed anatomy.

Homologous (Inferring Common Ancestry) Structures do not Provide Evidence for the Darwinian Model?

The *Campbell Biology* text[9] says, "A second type of evidence for evolution comes from analyzing similarities among different organisms. Characteristics present in an ancestral organism are altered (by natural selection) in its descendants over time as they face different environmental conditions. As a result, related species can have characteristics that have an underlying similarity yet function differently. Similarity resulting from common ancestry is known as homology." This popular text and the Louisiana Science standards are either unaware or intentionally neglecting to make students aware of what prominent scientists are saying about homology.[10] Once again, here is a list of scientists exposing the weaknesses in the Darwinian thinking.[11]

---

[9] Urry, Lisa A., et al. *Campbell Biology 11th Edition*. New York : Pearson, 2017, p. 477.

[10] For more discussion of the *Campbell Biology* text, please see chapter 6.

[11] See http://www.discovery.org/scripts/viewDB/filesDB-download.php?command=download&id=12175

1. **Michael Denton**, a famous author and biochemist once said, "The validity of the evolutionary interpretation of homology would have been greatly strengthened if embryological and genetic research could have shown that homologous structures were specified by homologous genes and followed homologous patterns of embryological development. Such homology would indeed by strongly suggestive of 'true relationship; of inheritance from a common ancestor'. But it has become clear that the principle cannot be extended in this way. Homologous structures are often specified by non-homologous genetic systems and the concept of homology can seldom be extended back into embryology."[12]

2. **Günter P. Wagner**, the Alison Richard Professor and Acting Chair of Ecology and Evolutionary Biology, Yale University, runs the Wagner Lab at Yale. In a 2007 publication, Wagner says, "Homology is an essential idea of biology, referring to the historical continuity of characters, but it is also conceptually highly elusive. The main difficulty is the apparently loose relationship between morphological characters and their genetic basis. Here I propose that it is the historical continuity of gene regulatory networks rather than the expression of individual homologous genes that underlies the homology of morphological characters."[13]

The standard definition of common ancestry says that all living things have descended (evolved) from one or at least a much smaller number of original life forms. The most common textbook evidence for common ancestry begins by defining homologies as characteristics inherited from "common ancestors."[14] Ultimately, the idea is to construct a tree of life that connects all living things to their evolutionary ancestry. These tree of

---

[12] Michael Denton, *Evolution: A Theory in Crisis* (Chevy Chase, MD: Adler and Adler, 1986), 145

[13] Günter P. Wagner, "*The developmental genetics of homology*," Perspectives, Nature Reviews Genetics 8 (June 2007): 473, doi:10.1038/nrg2099 Abstract.

[14] For a fuller description of how the *Campbell Biology* text presents this information, see chapter 6.

life presentations are the results of "phylogenetic analyses" and are often referred to as phylogenetic trees."

DNA Research does not Support Common Ancestry

As it turns out, the tree of life concept is in a state of confusion in the scientific literature. Scientists have known for years that these trees showing how organisms are related (evolutionarily) to their ancestors look very different depending on which genetic (DNA/RNA) sequence or phenotypic (morphological/anatomical) trait is being compared.[15] The language of the standard does not seem to be aware of this fact. As it turns out, conflict in the production of these trees is the norm and not the exception. This has led some scientists to conclude that the evolutionary tree concept depicting common ancestry is obsolete and should be discarded.[16] Ground breaking research in microRNAs led Dartmouth

---

[15] A) *Bones, Molecules or Both?* Gura, T. July 2000, Nature, Vol. 406, pp. 230-233. The complete mitochondrial DNA sequence of shark Mustelus manazo: Evaluating rooting contradictions to bony vertebrates. Ying Cao, Peter J. Waddell, Norhiro Okado, Masami hasegawa. 1998, Molecular Biology and Evolution, Vol. 15, pp. 1637-1646.

B) *18S gene trees are positively misleading for monocot/dicot phylogenetics.* Melvin R. Duvall, Autumn Bricker Ervin. 2004, Molecular Phylogenetics and Evolution, Vol. 30, pp. 97-106.

C) *Molecules vs. morphology in avian evolution: The case of the 'pelecaniform' birds.* S. Blair Hedges, Charles G. Sibley. October 1994, Proceedings of the National Academy of Sciences USA , Vol. 91, pp. 9861-9865.

D) *Conflicting phylogenetic signals at the base of the metazoan tree.* A. Rokas, N. King, J. Finnerty, S.B. Carroll. 2003, Evolution and Development, Vol. 5, pp. 346-359.

[16] A) *Understanding Phylogenetic Incongruence: Lessons from Phyllostomid Bats.* L.M. Dávalos, A.L. Cirranello, J.H. Geisler, N.B. Simmons. 4, 2012, Biological Reviews of the Cambridge Philosophical Society, Vol. 87, pp. 991-1024.

biologist Kevin Peterson to say, "I've looked at thousands of microRNA genes, and I can't find a single example that would support the traditional tree."[17] Presumably, only one real evolutionary tree can be correct. These conflicting results from researchers pose a serious problem to the Darwinian model.

Here's a list of well-known scientists and things they have said in the scientific literature about this problem.[18]

1. **Rudi Loesel** reported news from a conference titled, **Celebrating Darwin: From the Origin of Species to Deep Metazoan Phylogeny** which was held at the Humboldt University in Berlin in March 2009. Specialists from the fields of bioinformatics, molecular biology, developmental biology, comparative morphology and paleontology joined forces to present and discuss novel approaches in reconstructing the still unresolved early branching patterns of the metazoan tree of life. In part of the summary of the conference, the article notes: "Advance is truly needed because even 150 years after Darwin's On the Origin of Species, the origin of most major animal groups is still a matter of debate. With an average of 15 phylogenetic trees published per day the picture is not necessarily getting clearer. Many trees contradict each other."[19]

2. **W. Ford Doolittle**, Professor Emeritus, in the Department of Biochemistry and Molecular Biology at Dalhousie University, and Member of the United States National Academy of Sciences and **Eric Bapteste**, from the Department of Evolutionary Biology, Université Pierre et Marie

---

B) *Why Darwin Was Wrong about the Tree of Life*. Lawton, G. January 21, 2009, New Scientist, pp. 34-39. Phylogenetic Classification and the Universal Tree. Doolittle, W. F. June 25, 1999, Science, Vol. 284, pp. 2124-2128.

[17] *Rewriting Evolution*. Dolgin, E. June 28, 2012, Nature, Vol. 486, pp. 460-462.

[18] See http://www.discovery.org/scripts/viewDB/filesDB-download.php?command=download&id=12175

[19] From the abstract: "*150 years beyond Darwin's Origin of species: finding new approaches to reconstruct early animal phylogeny,*" Biology Letters 5 (May 14, 2009): 436-438; http://rsbl.royalsocietypublishing.org/content/5/4/436.

Curie, Paris, France are quoted as saying, "Darwin claimed that a unique inclusively hierarchical pattern of relationships between all organisms based on their similarities and differences [the Tree of Life (TOL)] was a fact of nature, for which evolution, and in particular a branching process of descent with modification, was the explanation. However, there is no independent evidence that the natural order is an inclusive hierarchy, and incorporation of prokaryotes into the TOL is especially problematic. The only data sets from which we might construct a universal hierarchy including prokaryotes, the sequences of genes, often disagree and can seldom be proven to agree. Hierarchical structure can always be imposed on or extracted from such data sets by algorithms designed to do so, but at its base the universal TOL rests on an unproven assumption about pattern that, given what we know about process, is unlikely to be broadly true."[20]

3. **S. Andrew Inkpen and W. Ford Doolittle** in a 2016 article state, "The concept of homology has a long history, during much of which the issue has been how to reconcile similarity and common descent when these are not coextensive. Although thinking molecular phylogeneticists have learned not to say 'percent homology,' the problems are deeper than that and unresolved."[21]

4. **Carl Woese** was the microbiologist who discovered domain Archaea and was one of the first to propose that RNA replication predates protein synthesis. In an article in the Proceedings of the National Academy of Sciences, he admits that there are phylogenetic incongruities throughout the tree of life, stating, "Phylogenetic incongruities can be seen everywhere in the universal tree, from its root to the major branchings within and

---

[20] W. Ford Doolittle and Eric Bapteste, *"Pattern pluralism and the Tree of Life hypothesis,"* Proceedings of the Biological Society of Washington USA 104, no. 7 (February 13, 2007): 2043-2049, http://www.pnas.org/content/104/7/2043.long.

[21] S. Andrew Inkpen, W. Ford Doolittle, *"Molecular Phylogenetics and the Perennial Problem of Homology,"* Journal of Molecular Evolution 83, no. 5-6 (November 21, 2016), doi: 10.1007/s00239-016-9766-4 Abstract.

among the various taxa to the makeup of the primary groupings themselves."[22]

5. **Vicky Merhej and Didier Raoult** published a notable article in 2012. Raoult has been called Europe's most cited microbiologist and is a Professor of Clinical Microbiology, Aix-Marseille Université. The article states, "Genomes are collections of genes with different evolutionary histories that cannot be represented by a single tree of life (TOL). A forest, a network or a rhizome of life may be more accurate to represent evolutionary relationships among species."[23]

6. **Maureen A. O'Malley** is from the Department of Philosophy, University of Sydney. **Eugene Koonin**, is an evolutionary molecular biologist, a Senior Investigator at the National Center for Biotechnology Information (National Library of Medicine, National Institutes of Health), is Editor-in-Chief of Biology Direct, and is a member of the National Academy of Sciences. In a 2011 article, these two say, "We examine the Tree of Life (TOL) as an evolutionary hypothesis and a heuristic. The original TOL hypothesis has failed but a new 'statistical TOL hypothesis' is promising. The TOL heuristic usefully organizes data without positing fundamental evolutionary truth."[24]

7. **David B. Wake, Marvalee H. Wake, and Chelsea D. Specht** published an important article stating, "Understanding the diversification of phenotypes through time—'descent with modification'— has been the

---

[22] Carl Woese, *"The universal ancestor,"* Proceedings of the National Academy of Sciences USA 95, no. 12 (June, 1998): 6854- 6859, http://www.pnas.org/content/95/12/6854.full.

[23] Vicky Merhej and Didier Raoult, *"Rhizome of life, catastrophes, sequence exchanges, gene creations, and giant viruses: how microbial genomics challenges,"* Frontiers in Cellular and Infection Microbiology 2 (August 28, 2012), https://www.ncbi.nlm.nih.gov/pmc/articles/PMC3428605/.

[24] Maureen A. O'Malley and Eugene V. Koonin, *"How stands the Tree of Life a century and a half after The Origin?"* Biology Direct 6, no. 32 (June 30, 2011), https://biologydirect.biomedcentral.com/articles/10.1186/1745-6150-6-32.

focus of evolutionary biology for 150 years. If, contrary to expectations, similarity evolves in unrelated taxa, researchers are guided to uncover the genetic and developmental mechanisms responsible. Similar phenotypes may be retained from common ancestry (homology), but a phylogenetic context may instead reveal that they are independently derived, due to convergence or parallel evolution, or less likely, that they experienced reversal. Such examples of homoplasy present opportunities to discover the foundations of morphological traits. A common underlying mechanism may exist, and components may have been redeployed in a way that produces the 'same' phenotype. New, robust phylogenetic hypotheses and molecular, genomic, and developmental techniques enable integrated exploration of the mechanisms by which similarity arises."[25]

8. **Biologist Emily Jane McTavish** and her team say in their 2017 article, "The question of whether a 'tree of life' exists has been discussed for decades, and the recent publication of genome level phylogenies across many taxa brings enormous quantities of empirical data to bear on this question. These data demonstrate that a single genome often contains regions with divergent evolutionary histories. Processes such as incomplete lineage sorting, horizontal gene transfer, endosymbioses, and the incorporation of viruses into genomes can drive differences in the shared evolutionary history in different parts of genomes."[26]

9. **Elizabeth Pennisi**, in a 2010 publication states, "It used to seem so straightforward. DNA told the body how to build proteins. The instructions came in chapters called genes. Strands of DNA's chemical cousin RNA served as molecular messengers, carrying orders to the cells' protein factories and translating them into action. Between the genes lay long stretches of 'junk DNA,' incoherent, useless, and inert. That was then. In fact, gene regulation has turned out to be a surprisingly complex process

---

[25] David B. Wake, Marvalee H. Wake, and Chelsea D. Specht, *"Homoplasy: From Detecting Pattern to Determining Process and Mechanism of Evolution,"* Science 331, no. 6020 (February 25, 2011): 1032-1035.

[26] Emily Jane McTavish et al., *"How and Why to Build a Unified Tree of Life,"* BioEssays 39, no. 11 (October 5, 2017): 1700114, doi: 10.1002/bies.201700114

governed by various types of regulatory DNA, which may lie deep in the wilderness of supposed 'junk.' Far from being humble messengers, RNAs of all shapes and sizes are actually powerful players in how genomes operate. Finally, there's been increasing recognition of the widespread role of chemical alterations called epigenetic factors that can influence the genome across generations without changing the DNA sequence itself."[27]

10. **Leanne S. Haggerty,** in a 2014 article in *Molecular Biology and Evolution* article says, "Defining homologous genes is important in many evolutionary studies but raises obvious issues....In particular, defining homologous genes cannot be solely addressed under the classic assumptions of strong tree thinking, according to which genes evolve in a strictly tree-like fashion of vertical descent and divergence and the problems of homology detection are primarily methodological. Gene homology could also be considered under a different perspective where genes evolve as "public goods," subjected to various introgressive processes. In this latter case, defining homologous genes becomes a matter of designing models suited to the actual complexity of the data and how such complexity arises, rather than trying to fit genetic data to some a priori tree-like evolutionary model, a practice that inevitably results in the loss of much information."[28]

---

[27] Elizabeth Pennisi, "*Shining a Light on the Genome's 'Dark Matter'*," Science 330, no. 6011 (December 17, 2010): 1614, http://science.sciencemag.org/content/330/6011/1614.full

[28] Leanne S. Haggerty, et al., "*A Pluralistic Account of Homology: Adapting the Models to the Data,*" Molecular Biology and Evolution 31, no. 3 (March 1, 2014): 501-516, doi:10.1093/molbev/mst228 Abstract.

CHAPTER FIVE

HIGH SCHOOL STANDARD 8-HS-LS4-1: COMMON ANCESTRY, DNA SEQUENCING, EMBRYOLOGY, VESTIGIAL STRUCTURES AND FOSSILS

PERFORMANCE EXPECTATION: ANALYZE AND INTERPRET scientific information that common ancestry and biological evolution are supported by multiple lines of empirical evidence.

Clarification Statement: Emphasis is on a conceptual understanding of the role each line of evidence (e.g., similarities in DNA sequences, order of appearance of structure during embryological development, cladograms, homologous and vestigial structures, fossil records) demonstrates as related to common ancestry and biological evolution.

Science and Engineering Practices: Analyzing and interpreting data: Analyzing data in 9-12 builds on K-8 experiences and progresses to introducing more detailed statistical analysis, the comparison of data sets for consistency, and the use of models to generate and analyze data. Compare and contrast various types of data sets (e.g., self-generated, archival) to examine consistency of measurements and observations.

Disciplinary Core Ideas: *EVIDENCE OF COMMON ANCESTRY AND DIVERSITY*. Genetic information provides evidence of evolution. DNA sequences vary among species, but there are many overlaps; in fact, the ongoing branching that produces multiple lines of descent can be inferred by comparing the DNA sequences of different organisms. Such information is also derivable from the similarities and differences in amino acid sequences and from observable anatomical and embryological evidence.

Crosscutting concepts: PATTERNS. Different patterns may be observed at each of the scales at which a system is studied and can provide evidence for causality in explanations of phenomena.

There are five standards in the high school section on "Biological Evolution: Unity and Diversity." This standard is found in the file called "Science Standards – Life Science,"[1] the standards for high school students. The content area of the standard to be addressed will be highlighted in bold italics. Much of the information in this chapter has been reproduced here for readers of the high school standards. Please notice the language of the standard to be discussed in bold italics.

Problem 1: The Fossil Record does not Support the Idea of Common Ancestry

Performance Expectation: Analyze and interpret scientific information that *common ancestry* and biological evolution are supported by multiple lines of empirical evidence.

Clarification Statement: Emphasis is on a conceptual understanding of the role each line of evidence (e.g., similarities in DNA sequences, order of appearance of structure during embryological development, cladograms, homologous and vestigial structures, fossil records) demonstrates as related to *common ancestry* and biological evolution.

Disciplinary Core Ideas: *EVIDENCE OF COMMON ANCESTRY AND DIVERSITY.* Genetic information provides evidence of evolution. DNA sequences vary among species, but there are many overlaps; in fact, the ongoing branching that produces multiple lines of descent can be inferred by comparing the DNA sequences of different organisms. Such information is also derivable from the similarities and differences in amino acid sequences and from observable anatomical and embryological evidence.

The standard definition of common ancestry says that all living things have descended (evolved) from one or at least a much smaller number of original life forms. The most common textbook evidence for common ancestry begins by defining homologies as characteristics inherited from

---

[1] See https://www.louisianabelieves.com/resources/library/academic-standards

"common ancestors."[2] Ultimately, the idea is to construct a tree of life that connects all living things to their evolutionary ancestry. These tree of life presentations are the results of "phylogenetic analyses" and are often referred to as "phylogenetic trees."

The *Campbell Biology* text says, "A second type of evidence for evolution comes from analyzing similarities among different organisms. Characteristics present in an ancestral organism are altered (by natural selection) in its descendants over time as they face different environmental conditions. As a result, related species can have characteristics that have an underlying similarity yet function differently. Similarity resulting from common ancestry is known as **homology**." This popular text and the Louisiana Science standards are either unaware or intentionally neglecting to make students aware of what prominent scientists are saying about homology.[3] Once again, here is a list of scientists exposing the weaknesses in the Darwinian thinking, this time on homologies.[4]

1. **Michael Denton**, a famous author and biochemist once said, "The validity of the evolutionary interpretation of homology would have been greatly strengthened if embryological and genetic research could have shown that homologous structures were specified by homologous genes and followed homologous patterns of embryological development. Such homology would indeed by strongly suggestive of 'true relationship; of inheritance from a common ancestor'. But it has become clear that the principle cannot be extended in this way. Homologous structures are often specified by non-homologous genetic systems and the concept of homology can seldom be extended back into embryology."[5]

---

[2] For a fuller description of how the *Campbell Biology* text presents this information, see chapter 6.

[3] For more discussion of the *Campbell Biology* text, please see chapter 6.

[4] See http://www.discovery.org/scripts/viewDB/filesDB-download.php?command=download&id=12175

[5] Michael Denton, *Evolution: A Theory in Crisis* (Chevy Chase, MD: Adler and Adler, 1986), 145

2. **Günter P. Wagner**, the Alison Richard Professor and Acting Chair of Ecology and Evolutionary Biology, Yale University, runs the Wagner Lab at Yale. In a 2007 publication, Wagner says, "Homology is an essential idea of biology, referring to the historical continuity of characters, but it is also conceptually highly elusive. The main difficulty is the apparently loose relationship between morphological characters and their genetic basis. Here I propose that it is the historical continuity of gene regulatory networks rather than the expression of individual homologous genes that underlies the homology of morphological characters."[6]

The standard definition of common ancestry says that all living things have descended (evolved) from one or at least a much smaller number of original life forms. The most common textbook evidence for common ancestry begins by defining homologies as characteristics inherited from "common ancestors."[7] Ultimately, the idea is to construct a tree of life that connects all living things to their evolutionary ancestry. These tree of life presentations are the results of "phylogenetic analyses" and are often referred to as phylogenetic trees."

Problem 2: DNA Sequence Comparisons Present Major Problems for Determining Common Ancestry

Clarification Statement: Emphasis is on a conceptual understanding of the role each line of evidence (e.g., *similarities in DNA sequences*, order of appearance of structure during embryological development, cladograms, homologous and vestigial structures, fossil records) demonstrates as related to common ancestry and biological evolution.

Disciplinary Core Ideas: *EVIDENCE OF COMMON ANCESTRY AND DIVERSITY.* Genetic information provides evidence of evolution. DNA sequences vary among species, but there are many overlaps; in fact, the ongoing branching that produces multiple lines

---

[6] Günter P. Wagner, "*The developmental genetics of homology*," Perspectives, Nature Reviews Genetics 8 (June 2007): 473, doi:10.1038/nrg2099 Abstract.

[7] For a fuller description of how the *Campbell Biology* text presents this information, see chapter 6.

of descent can be inferred by *comparing the DNA sequences of different organisms*. Such information is also derivable from the similarities and differences in amino acid sequences and from observable anatomical and embryological evidence.

Remember that the standard definition of common ancestry says that all living things have descended (evolved) from one or at least a much smaller number of original life forms. The most common textbook evidence for common ancestry begins by defining homologies as characteristics inherited from "common ancestors."[8] These characteristics are what are referred to in the clarification statement of the standard as *"similarities in DNA sequences, order of appearance of structure during embryological development, cladograms, homologous and vestigial structures."* Ultimately, the idea is to construct a tree of life that connects all living things to their ancestry. These tree of life presentations are the results of "phylogenetic analyses" and are often referred to as "phylogenetic trees."

DNA Research does not Support Common Ancestry

As it turns out, the tree of life concept is in a state of confusion in the scientific literature. Scientists have known for years that these trees showing how organisms are related (evolutionarily) to their ancestors look very different depending on which genetic (DNA/RNA) sequence or phenotypic (morphological/anatomical) trait is being compared.[9] The

---

[8] For a fuller description of how the *Campbell Biology* text presents this information, see chapter 6.

[9] A) *Bones, Molecules or Both?* Gura, T. July 2000, Nature, Vol. 406, pp. 230-233. The complete mitochondrial DNA sequence of shark Mustelus manazo: Evaluating rooting contradictions to bony vertebrates. Ying Cao, Peter J. Waddell, Norhiro Okado, Masami hasegawa. 1998, Molecular Biology and Evolution, Vol. 15, pp. 1637-1646.

B) *18S gene trees are positively misleading for monocot/dicot phylogenetics*. Melvin R. Duvall, Autumn Bricker Ervin. 2004, Molecular Phylogenetics and Evolution, Vol. 30, pp. 97-106.

language of the standard does not seem to be aware of this fact. As it turns out, conflict in the production of these trees is the norm and not the exception. This has led some scientists to conclude that the evolutionary tree concept depicting common ancestry is obsolete and should be discarded.[10] Ground breaking research in microRNAs led Dartmouth biologist Kevin Peterson to say, "I've looked at thousands of microRNA genes, and I can't find a single example that would support the traditional tree."[11] Presumably, only one real evolutionary tree can be correct. These conflicting results from researchers pose a serious problem to the Darwinian model.

Here's a list of well-known scientists and things they have said in the scientific literature about this problem.[12]

1. **Rudi Loesel** reported news from a conference titled, **Celebrating Darwin: From the Origin of Species to Deep Metazoan Phylogeny** which was held at the Humboldt University in Berlin in March 2009. Specialists from the fields of bioinformatics, molecular biology, developmental biology, comparative morphology and paleontology joined

---

C) *Molecules vs. morphology in avian evolution: The case of the 'pelecaniform' birds.* S. Blair Hedges, Charles G. Sibley. October 1994, Proceedings of the National Academy of Sciences USA , Vol. 91, pp. 9861-9865.

D) *Conflicting phylogenetic signals at the base of the metazoan tree.* A. Rokas, N. King, J. Finnerty, S.B. Carroll. 2003, Evolution and Development, Vol. 5, pp. 346-359.

[10] A) *Understanding Phylogenetic Incongruence: Lessons from Phyllostomid Bats.* L.M. Dávalos, A.L. Cirranello, J.H. Geisler, N.B. Simmons. 4, 2012, Biological Reviews of the Cambridge Philosophical Society, Vol. 87, pp. 991-1024.

B) *Why Darwin Was Wrong about the Tree of Life.* Lawton, G. January 21, 2009, New Scientist, pp. 34-39. Phylogenetic Classification and the Universal Tree. Doolittle, W. F. June 25, 1999, Science, Vol. 284, pp. 2124-2128.

[11] *Rewriting Evolution.* Dolgin, E. June 28, 2012, Nature, Vol. 486, pp. 460-462.

[12] See http://www.discovery.org/scripts/viewDB/filesDB-download.php?command=download&id=12175

forces to present and discuss novel approaches in reconstructing the still unresolved early branching patterns of the metazoan tree of life. In part of the summary of the conference, the article notes: "Advance is truly needed because even 150 years after Darwin's On the Origin of Species, the origin of most major animal groups is still a matter of debate. With an average of 15 phylogenetic trees published per day the picture is not necessarily getting clearer. Many trees contradict each other."[13]

2. **W. Ford Doolittle**, Professor Emeritus, in the Department of Biochemistry and Molecular Biology at Dalhousie University, and Member of the United States National Academy of Sciences and **Eric Bapteste**, from the Department of Evolutionary Biology, Université Pierre et Marie Curie, Paris, France are quoted as saying, "Darwin claimed that a unique inclusively hierarchical pattern of relationships between all organisms based on their similarities and differences [the Tree of Life (TOL)] was a fact of nature, for which evolution, and in particular a branching process of descent with modification, was the explanation. However, there is no independent evidence that the natural order is an inclusive hierarchy, and incorporation of prokaryotes into the TOL is especially problematic. The only data sets from which we might construct a universal hierarchy including prokaryotes, the sequences of genes, often disagree and can seldom be proven to agree. Hierarchical structure can always be imposed on or extracted from such data sets by algorithms designed to do so, but at its base the universal TOL rests on an unproven assumption about pattern that, given what we know about process, is unlikely to be broadly true."[14]

3. **S. Andrew Inkpen and W. Ford Doolittle** in a 2016 article state, "The concept of homology has a long history, during much of which the issue has been how to reconcile similarity and common descent when these

---

[13] From the abstract: *"150 years beyond Darwin's Origin of species: finding new approaches to reconstruct early animal phylogeny,"* Biology Letters 5 (May 14, 2009): 436-438; http://rsbl.royalsocietypublishing.org/content/5/4/436.

[14] W. Ford Doolittle and Eric Bapteste, *"Pattern pluralism and the Tree of Life hypothesis,"* Proceedings of the Biological Society of Washington USA 104, no. 7 (February 13, 2007): 2043-2049, http://www.pnas.org/content/104/7/2043.long.

are not coextensive. Although thinking molecular phylogeneticists have learned not to say 'percent homology,' the problems are deeper than that and unresolved."[15]

4. **Carl Woese** was the microbiologist who discovered domain Archaea and was one of the first to propose that RNA replication predates protein synthesis. In an article in the Proceedings of the National Academy of Sciences, he admits that there are phylogenetic incongruities throughout the tree of life, stating, "Phylogenetic incongruities can be seen everywhere in the universal tree, from its root to the major branchings within and among the various taxa to the makeup of the primary groupings themselves."[16]

5. **Vicky Merhej and Didier Raoult** published a notable article in 2012. Raoult has been called Europe's most cited microbiologist and is a Professor of Clinical Microbiology, Aix-Marseille Université. The article states, "Genomes are collections of genes with different evolutionary histories that cannot be represented by a single tree of life (TOL). A forest, a network or a rhizome of life may be more accurate to represent evolutionary relationships among species."[17]

6. **Maureen A. O'Malley** is from the Department of Philosophy, University of Sydney. **Eugene Koonin**, is an evolutionary molecular biologist, a Senior Investigator at the National Center for Biotechnology Information (National Library of Medicine, National Institutes of Health),

---

[15] S. Andrew Inkpen, W. Ford Doolittle, *"Molecular Phylogenetics and the Perennial Problem of Homology,"* Journal of Molecular Evolution 83, no. 5-6 (November 21, 2016), doi: 10.1007/s00239-016-9766-4 Abstract.

[16] Carl Woese, *"The universal ancestor,"* Proceedings of the National Academy of Sciences USA 95, no. 12 (June, 1998): 6854- 6859, http://www.pnas.org/content/95/12/6854.full.

[17] Vicky Merhej and Didier Raoult, *"Rhizome of life, catastrophes, sequence exchanges, gene creations, and giant viruses: how microbial genomics challenges,"* Frontiers in Cellular and Infection Microbiology 2 (August 28, 2012), https://www.ncbi.nlm.nih.gov/pmc/articles/PMC3428605/.

is Editor-in-Chief of Biology Direct, and is a member of the National Academy of Sciences. In a 2011 article, these two say, "We examine the Tree of Life (TOL) as an evolutionary hypothesis and a heuristic. The original TOL hypothesis has failed but a new 'statistical TOL hypothesis' is promising. The TOL heuristic usefully organizes data without positing fundamental evolutionary truth."[18]

7. **David B. Wake, Marvalee H. Wake, and Chelsea D. Specht** published an important article stating, "Understanding the diversification of phenotypes through time—'descent with modification'— has been the focus of evolutionary biology for 150 years. If, contrary to expectations, similarity evolves in unrelated taxa, researchers are guided to uncover the genetic and developmental mechanisms responsible. Similar phenotypes may be retained from common ancestry (homology), but a phylogenetic context may instead reveal that they are independently derived, due to convergence or parallel evolution, or less likely, that they experienced reversal. Such examples of homoplasy present opportunities to discover the foundations of morphological traits. A common underlying mechanism may exist, and components may have been redeployed in a way that produces the 'same' phenotype. New, robust phylogenetic hypotheses and molecular, genomic, and developmental techniques enable integrated exploration of the mechanisms by which similarity arises."[19]

8. **Biologist Emily Jane McTavish** and her team say in their 2017 article, "The question of whether a 'tree of life' exists has been discussed for decades, and the recent publication of genome level phylogenies across many taxa brings enormous quantities of empirical data to bear on this question. These data demonstrate that a single genome often contains regions with divergent evolutionary histories. Processes such as incomplete

---

[18] Maureen A. O'Malley and Eugene V. Koonin, "*How stands the Tree of Life a century and a half after The Origin?*" Biology Direct 6, no. 32 (June 30, 2011), https://biologydirect.biomedcentral.com/articles/10.1186/1745-6150-6-32.

[19] David B. Wake, Marvalee H. Wake, and Chelsea D. Specht, *"Homoplasy: From Detecting Pattern to Determining Process and Mechanism of Evolution,"* Science 331, no. 6020 (February 25, 2011): 1032-1035.

lineage sorting, horizontal gene transfer, endosymbioses, and the incorporation of viruses into genomes can drive differences in the shared evolutionary history in different parts of genomes."[20]

9. **Elizabeth Pennisi**, in a 2010 publication states, "It used to seem so straightforward. DNA told the body how to build proteins. The instructions came in chapters called genes. Strands of DNA's chemical cousin RNA served as molecular messengers, carrying orders to the cells' protein factories and translating them into action. Between the genes lay long stretches of 'junk DNA,' incoherent, useless, and inert. That was then. In fact, gene regulation has turned out to be a surprisingly complex process governed by various types of regulatory DNA, which may lie deep in the wilderness of supposed 'junk.' Far from being humble messengers, RNAs of all shapes and sizes are actually powerful players in how genomes operate. Finally, there's been increasing recognition of the widespread role of chemical alterations called epigenetic factors that can influence the genome across generations without changing the DNA sequence itself."[21]

10. **Leanne S. Haggerty,** in a 2014 article in *Molecular Biology and Evolution* article says, "Defining homologous genes is important in many evolutionary studies but raises obvious issues....In particular, defining homologous genes cannot be solely addressed under the classic assumptions of strong tree thinking, according to which genes evolve in a strictly tree-like fashion of vertical descent and divergence and the problems of homology detection are primarily methodological. Gene homology could also be considered under a different perspective where genes evolve as "public goods," subjected to various introgressive processes. In this latter case, defining homologous genes becomes a matter of designing models suited to the actual complexity of the data and how such complexity arises,

---

[20] Emily Jane McTavish et al., *"How and Why to Build a Unified Tree of Life,"* BioEssays 39, no. 11 (October 5, 2017): 1700114, doi: 10.1002/bies.201700114

[21] Elizabeth Pennisi, *"Shining a Light on the Genome's 'Dark Matter',"* Science 330, no. 6011 (December 17, 2010): 1614, http://science.sciencemag.org/content/330/6011/1614.full

rather than trying to fit genetic data to some a priori tree-like evolutionary model, a practice that inevitably results in the loss of much information."[22]

Problem 3: Embryological Evidence does not Support the Darwinian Model

Clarification Statement: Emphasis is on a conceptual understanding of the role each line of evidence (e.g., similarities in DNA sequences, ***order of appearance of structure during embryological development***, cladograms, homologous and vestigial structures, fossil records) demonstrates as related to common ancestry and biological evolution.

Embryology, simply stated, is the study of an growing embryo. The *Campbell Biology* text[23] states that, "At some stage in their embryonic development, all vertebrates have a tail located posterior to the anus (referred to as a post-anal tail), as well as pharyngeal (throat) arches. Descent from a common ancestor can explain such similarities." This argument was a favorite of Darwin's and made famous by Ernst Haeckel in the late 1800s.

As it turns out, much controversy clouds the use of embryological drawings to suggest relatedness between species.[24] The real question is how much of this controversy should the students know about. My answer would be to tell them the entire story. It's a great illustration of bias and corruption that exists in the scientific community.

---

[22] Leanne S. Haggerty, et al., *"A Pluralistic Account of Homology: Adapting the Models to the Data,"* Molecular Biology and Evolution 31, no. 3 (March 1, 2014): 501-516, doi:10.1093/molbev/mst228 Abstract.

[23] Urry, Lisa A., et al. *Campbell Biology 11th Edition.* New York : Pearson, 2017, p. 477.

[24] For a fuller explanation, see chapter 6.

Here are comments of well-known scientists accompanied by the citations in the footnotes. Once again internet addresses are provided where possible to aid in easy access.[25]

1. **Jonathan Wells**, in an article about Haeckel' embryos, is quoted as saying, "In The Origin of Species Charles Darwin wrote that 'the embryos of mammals, birds, fishes, and reptiles [are] closely similar, but become, when fully developed, widely dissimilar.' He inferred that all vertebrates 'are the modified descendants of some ancient progenitor,' and that 'the embryonic or larval stages show us, more or less completely, the condition of the progenitor of the whole group in its adult state' (Darwin 1859, pp. 338, 345). Darwin's contemporary Ernst Haeckel called this the 'Biogenetic Law,' according to which 'ontogeny recapitulates phylogeny.' To illustrate the law, Haeckel (1891) produced drawings of vertebrate embryos which have been widely used in biology textbooks ever since (Figure 1). "But Haeckel's Biogenetic Law was discredited by embryologists in Darwin's lifetime (Bowler 1989); recent work has shown that Haeckel's drawings misrepresent the embryos they purport to show (Richardson et al. 1997); and Haeckel entirely omitted the earliest stages of development in which the various classes of vertebrates are morphologically very different (Elinson 1987). Biology teachers should be aware that Haeckel's drawings do not fit the facts."[26]

2. **Stephen J. Gould** (1941 – 2002) was a Harvard University evolutionary biologist. He said, "We should therefore not be surprised that Haeckel's drawings entered nineteenth-century textbooks. But we do, I think, have the right to be both astonished and ashamed by the century of mindless recycling that has led to the persistence of these drawings in a large number, if not a majority, of modern textbooks!"[27]

---

[25] Reprinted with permission from the Discovery Institute and can be found at: http://www.discovery.org/scripts/viewDB/filesDB-download.php?command=download&id=12175

[26] J. Wells, *"Haeckel's Embryos & Evolution: Setting the Record Straight,"* American Biology Teacher 61, no. 5 (May 1999): 345-349.

[27] Stephen Jay Gould, *"Abscheulich! (Atrocious!),"* Natural History (March 2000).

3. **Elizabeth Pennisi** has written an article acknowledging that Haeckel's embryo drawings are both fraudulent and continue to be used in textbooks. It states: "generations of biology students may have been misled by a famous set of drawings of embryos published 123 years ago by the German biologist Ernst Haeckel. They show vertebrate embryos of different animals passing through identical stages of development. But the impression they give, that the embryos are exactly alike, is wrong." The article quotes further quotes leading embryologist **Michael Richardson** by stating that "it looks like it's turning out to be one of the most famous fakes in biology."[28]

4. **Alex T. Kalinka, Karolina M. Varga et al**. Both Kalinka and Varga are of the Max Planck Institute of Molecular Cell Biology and Genetics. This article challenges claims of early similarities between embryos. Many textbooks simply show images of two vertebrate embryos and ask the reader to imagine that since they look similar common descent should be considered as a viable concept. These researchers looked at genes from several different species of Drosophila and found that these genes reflected the hourglass model. And some types of genes adhered more closely than others.[29] The hourglass model imagines that both early and late embryonic forms are more diverse than the intermediate stages of development.

Problem 4: Vestigial Structures do not Support the Darwinian Model

Clarification Statement: Emphasis is on a conceptual understanding of the role each line of evidence (e.g., similarities in DNA sequences, order of appearance of structure during embryological development, cladograms,

---

[28] Elizabeth Pennisi, *"Haeckel's Embryos: Fraud Rediscovered,"* Science 277, no. 5331 (September 5, 1997): 1435a.

[29] Alex T Kalinka, Karolina M Varga et al., *"Gene expression divergence recapitulates the developmental hourglass model,"* Nature 468 (December 9, 2010): 811.

homologous and *vestigial structures*, fossil records) demonstrates as related to common ancestry and biological evolution.

Are They Really Vestigial?

The argument goes something like this for vestigial structures. These structures have seemingly lost their original function and perhaps have no function at this point in the evolutionary history of the organism as a result of random evolutionary processes. Not so fast, some prominent scientists would say.[30]

1. **Charles Q. Choi** quotes researcher William Parker, an immunologist at Duke University Medical Center in Durham, N.C., in a 2009 article. According to Choi, Parker has said, "Maybe it's time to correct the textbooks. Many biology texts today still refer to the appendix as a 'vestigial organ.'"[31]

2. **Heather F. Smith** is an Associate Professor of Anatomy at Midwestern University. **William Parker** is an Associate Professor of Surgery at Duke University School of Medicine. These two along with **Sanet H. Kotzé,** and **Michel Laurin** wrote that, "Although the cecal appendix has been widely viewed as a vestige with no known function or a remnant of a formerly utilized digestive organ, the evolutionary history of this anatomical structure is currently unresolved. A database was compiled for 361 mammalian species, and appendix characters were mapped onto a consensus phylogeny along with other gastrointestinal and behavioral characters. No correlation was found between appearance of an appendix and evolutionary changes in diet, fermentation strategy, coprophagia, social group size, activity pattern, cecal shape, or colonic separation mechanism. Appendix presence and size are positively correlated with cecum and colon size, even though this relationship rests largely on the larger size of cecum

---

[30] See http://www.discovery.org/scripts/viewDB/filesDB-download.php?command=download&id=12175

[31] Charles Q. Choi, *"The Appendix: Useful and in Fact Promising,"* LiveScience, August 24, 2009. https://www.livescience.com/10571-appendix-fact-promising.html.

and colon in taxa that have an appendix. The appendix has evolved minimally 32 times, but was lost fewer than seven times, indicating that it either has a positive fitness value or is closely associated with another character that does. These results, together with immunological and medical evidence, refute some of Darwin's hypotheses and suggest that the appendix is adaptive but has not evolved as a response to any particular dietary or social factor evaluated here."[32]

3. **Loren G. Martin**, a Professor of Physiology at Oklahoma State University says in a 2017 article, "For years, the appendix was credited with very little physiological function. We now know, however, that the appendix serves an important role in the fetus and in young adults. Endocrine cells appear in the appendix of the human fetus at around the 11th week of development. These endocrine cells of the fetal appendix have been shown to produce various biogenic amines and peptide hormones, compounds that assist with various biological control (homeostatic) mechanisms... Among adult humans, the appendix is now thought to be involved primarily in immune functions."[33]

4. The authors of the ***Inquiry Into Life*** textbook discuss the function of the tonsils, previously thought to be vestigial. "Fewer tonsillectomies are performed today than in the past because it is now known that the tonsils remove many of the pathogens that enter the pharynx; therefore, they are a first line of defense against invasion of the body."[34]

---

[32] Heather F. Smith, William Parker, Sanet H. Kotzé, and Michel Laurin, *"Multiple independent appearances of the cecal appendix in mammalian evolution and an investigation of related ecological and anatomical factors,"* Comptes Rendus Palevol 12, no. 6 (2013): 339-354, http://dx.doi.org/10.1016/j.crpv.2012.12.001 Abstract.

[33] "What is the function of the human appendix? Did it once have a purpose that has since been lost?" Response from Loren G. Martin, Scientific American, accessed November 2, 2017, https://www.scientificamerican.com/article/what-is-the-function-ofthe-human-appendix-did-it-once-have-a-purpose-that-has-since-been-lost/.

[34] Sylvia S. Mader, *Inquiry into Life*, 10th ed. (McGraw Hill, 2003), 293.

6. **Ting Jia** and **Eric G. Pamer**, the Head of the Division of Subspecialty Medicine at Memorial Sloan Kettering Cancer Center discuss the function of the spleen, previously thought to be vestigial in a 2009 paper. They state that, "In the left upper quadrant of the abdomen lies the spleen, functioning in two major capacities—filtering and storing blood cells, and acting as an immune tissue, where antibody synthesis occurs and certain pathogens are eliminated. Yet the spleen lacks the gravitas of neighboring organs because we can survive without it, albeit with some inconveniences. Its surgical removal causes modest increases in circulating white blood cells and platelets, diminished responsiveness to certain vaccines, and increased susceptibility to infection with certain bacteria and protozoa. But on page 612 in this issue, the organ gains some new respect, as Swirski et al. show that in the mouse, the spleen serves as a reservoir for immune cells (monocytes) that function in repairing the heart after myocardial infarction."[35]

7. **National Geographic News**. "'History is littered with body parts that were called 'useless' simply because medical science had yet to understand them,' [Jeffrey] Laitman [president-elect at the American Association of Anatomists] said. 'People say, You can remove it and still live. But you have to be careful with that logic,' he said. 'You could remove your left leg and still live. But whenever a body part is moved or changed, there's a price to pay.'"[36]

---

[35] Ting Jia and Eric G. Pamer, *"Immunology: Dispensable But Not Irrelevant,"* Science 325, no. 5940 (July 31 2009): 549-550, https://www.ncbi.nlm.nih.gov/pmc/articles/PMC2917045/

[36] Maggie Koerth-Baker, *"Vestigial Organs Not So Useless After All, Studies Find,"* National Geographic News, July 30, 2009, https://news.nationalgeographic.com/news/2009/07/090730-spleen-vestigial-organs.html.

Problem 5: Abrupt Appearance in the Fossil Record is not Expected in Modeling of Biological Evolution

Clarification Statement: Emphasis is on a conceptual understanding of the role each line of evidence (e.g., similarities in DNA sequences, order of appearance of structure during embryological development, cladograms, homologous and vestigial structures, *fossil records*) demonstrates as related to common ancestry and biological evolution.

All biology texts that discuss the Cambrian fossils, teach the concept of **abrupt appearance**. For example, the *Campbell Biology* text, one of the most popular texts in the country[37], discusses the Cambrian fossils by saying, "Many present-day animal phyla appear suddenly in fossils formed 535-525 million years ago early in the Cambrian period. This phenomenon is referred to as the Cambrian explosion. It goes on to say, "In a relatively short period of time, predators over 1 meter in length emerged that had claws and other features for capturing prey; simultaneously new defensive adaptations such as sharp spines and heavy body armor, appeared in their prey."[38]

The authors of the *Cambell Biology* text openly promote Darwinian evolution. In the section titled, "Our Hallmark Features," of the preface of the text, the authors say, "Chief among the themes of both Vision and Change and *Campbell's Biology* is evolution. Each chapter of this text includes at least one Evolution section that explicitly focuses on evolutionary aspects of the chapter material, and each chapter ends with an Evolution Connection Question."[39] This text does not suggest that this is a fatal flaw in the Darwinian model. However, students should have the opportunity to discuss the implications of fossil research that indicates that many (some say most) of the known animal phyla appear in Cambrian fossil

---

[37] Pearson. *Campbell Biology 11th edition*. Pearson. [Online] 2017. [Cited: December 31, 2017.] https://www.pearson.com/us/higher-education/program/Urry-Campbell-Biology-Plus-Mastering-Biology-with-Pearson-e-Text-Access-Card-Package-11th-Edition/PGM209351.html.

[38] Urry, Lisa A., et al. *Campbell Biology 11th Edition*. New York : Pearson, 2017.

[39] For a fuller discussion of the *Cambell's Biology* text, please see chapter 6.

beds without evidence of their Darwinian model ancestors. The next chapter address specifics from the *Campbell Biology* text.

The problems that research in the fossil beds have uncovered are not hard to find in the scientific literature. Please find below "Problem 6," a list of scientists with identifying descriptions who have published work on this topic.[40] Narrative explanations follow each, with quotation(s). The references can be found in the footnotes. Where possible, internet addresses are in the footnotes to help teachers access the information quickly.

Problem 6: The Evidence from the Fossil Record does not Support the Darwinian Model

Clarification Statement: Emphasis is on a conceptual understanding of the role each line of evidence (e.g., similarities in DNA sequences, order of appearance of structure during embryological development, cladograms, homologous and vestigial structures, ***fossil records***) demonstrates as related to common ancestry and biological evolution.

Weaknesses in the Darwinian Model - Fossils

1. **Stephen J. Gould** (1941-2002) was a famous evolutionary biologist who taught at Harvard University and was a member of the National Academy of Sciences. A classic 1972 paper by Gould argued that the history of fossil species includes two features inconsistent with Darwinian gradualism: 1) Stasis—that is most species exhibit no directional change during their tenure on earth, and 2) Sudden appearance.[41] A few years later

---

[40] Reprinted with permission from the Discovery Institute and can be found at: http://www.discovery.org/scripts/viewDB/filesDB-download.php?command=download&id=12175

[41] "*Punctuated Equilibria, an Alternative to Phyletic Gradualism*," Models in Paleobiology (1972): 82-115,
https://books.google.com/books?hl=en&lr=&id=3ULyAgAAQBAJ&oi=fnd&pg=PA239&dq=punctuated+equilibria+an+alternative+to+phyletic+gradualism+&ots=j_9-

Gould remarked that "The extreme rarity of transitional forms in the fossil record persists as the trade secret of paleontology. The evolutionary trees that adorn our textbooks have data only at the tips and nodes of their branches; the rest is inference, however reasonable, not the evidence of fossils.[42]

2. **Ernst Mayr** (1904-2005) was an evolutionary biologist from Harvard University; and a member of the National Academy of Sciences. Considered one of the greatest evolutionary biologists of the 20th century, Mayr noted, "Wherever we look at the living biota ... discontinuities are overwhelmingly frequent... The discontinuities are even more striking in the fossil record. New species usually appear in the fossil record suddenly, not connected with their ancestors by a series of intermediates."[43]

3. **Gareth Nelson** was a former vertebrate zoology curator at the American Museum for Natural History. "The idea that one can go to the fossil record and expect to empirically recover an ancestor-descendant sequence, be it of species, genera, families, or whatever, has been, and

---

FjBqn&sig=aDYDor5cC4wlpQvqJX6YeR0OsTA#v=onepage&q=punctuated%20equilibria%20an%20alternative%20to%20phyle tic%20gradualism&f=false

[42] Stephen J. Gould, "Evolution's Erratic Pace," Natural History 86, no. 5 (1977): 12–16, https://www.researchgate.net/publication/234661562_Evolution%27s_Erratic_Pace.

[43] Ernst Mayr, *What Evolution Is* (New York: Basic Books, 2001), 189, https://books.google.com/books?id=Dtg3l5QTAPsC&pg=PA189&dq=species+usually+appear+in+the+fossil+record+suddenly,+not+connected+with+their+ancestors+by+a+series+of+intermediates&hl=en&sa=X&ved=0ahUKEwjz_ZXIm5bVAhVW0GMKHTNBBuEQ6AEIMDAC#v=onepage&q=species%20usually%20appear%20in%20the%20fossil%20record%20suddenly%2C%20not%20connected%20with%20their%20ancestors%20by%20a%20series%20of%20intermediates&f=false.

continues to be, a pernicious illusion."[44] The phrase "the fossil record sounds impressive and authoritative." As used by some persons it becomes, as intended, intimidating, taking on the aura of esoteric truth as expounded by an elite class of specialists. But what is it, really, this fossil record? Only data in search of interpretation. All claims to the contrary that I know, and I know of several, are so much superstition.[45]

4. **David M. Raup**, is a former curator of Geology at the Field Museum of Natural History said, "Darwin ... was embarrassed by the fossil record ... Well, we are now about 120 years after Darwin and the knowledge of the fossil record has been greatly expanded. We now have a quarter of a million fossil species but the situation hasn't changed much. The record of evolution is still surprisingly jerky and, ironically, we have even fewer examples of evolutionary transitions than we had in Darwin's time. By this I mean that some of the classic cases of Darwinian change in the fossil record, such as the evolution of the horse in North America, have had to be discarded or modified as a result of more detailed information -- what appeared to be a nice simple progression when relatively few data were available now appear to be much more complex and much less gradualistic. So Darwin's problem has not been alleviated in the last 120 years and we still have a record which does show change but one that can hardly be looked upon as the most reasonable consequence of natural selection."[46]

---

[44] From a presentation by Gareth Nelson in 1969 to the American Museum of Natural History, quoted on page 709 in: David M. Williams and Malte C. Ebach, "The Reform of Palaeontology and the Rise of Biogeography — 25 Years after 'Ontogeny, Phylogeny, Palaeontology and the Biogenetic Law' (Nelson, 1978)," Journal of Biogeography 31, no. 5 (2004): 685–712, http://onlinelibrary.wiley.com/doi/10.1111/j.1365-2699.2004.01063.x/abstract.

[45] Gareth Nelson, *"Ontogeny, Phylogeny, Palaeontology and the Biogenetic Law,"* Systematic Zoology 27 (1978): 324–345, https://www.researchgate.net/publication/31252913_Ontogeny_Phylogeny_Paleontology_and_the_Biogenetic_Law.

[46] David M. Raup, "*Conflicts between Darwin and paleontology*," Field Museum of Natural History Bulletin 50 (1979): 22–29, https://archive.org/details/cbarchive_121465_conflictsbetweendarwinandpaleo1930.

5. **Henry Gee** is a vertebrate paleontologist and senior editor for biology at Nature. "We call new fossil discoveries missing links as if the chain of ancestry and descent were a real object for our contemplation, and not what it really is: a completely human invention created after the fact, shaped to accord with human prejudices... No fossil is buried with its birth certificate. That, and the scarcity of fossils, means that it is effectively impossible to link fossils into chains of cause and effect in any valid way... To take a line of fossils and claim that they represent a lineage is not a scientific hypothesis that can be tested, but an assertion that carries the same validity as a bedtime story—amusing, perhaps even instructive, but not scientific."[47]

6. **Eugene Koonin** is an evolutionary molecular biologist and a Senior Investigator for the National Center for Biotechnology Information (National Library of Medicine, National Institutes of Health), Editor-in-Chief of Biology Direct, and a member of the National Academy of Sciences. He has said, "Major transitions in biological evolution show the same pattern of sudden emergence of diverse forms at a new level of complexity. The relationships between major groups within an emergent new class of biological entities are hard to decipher and do not seem to fit the tree pattern that, following Darwin's original proposal, remains the dominant description of biological evolution."[48]

7. **Christian Schwabe** is a Professor of Biochemistry and Molecular Biology at Medical University of South Carolina. Schwabe provides a

---

[47] Henry Gee, *In Search of Deep Time: Going Beyond the Fossil Record to a Revolutionary Understanding of the History of Life* (London and Ithaca, NY; Comstock Publishing Associates, 1999), 32, 113, 116, https://books.google.de/books?id=TInB03o5uegC&printsec=frontcover&hl=de#v=onepage&q&f=false.

[48] Eugene V. Koonin, *"The Biological Big Bang model for the major transitions in evolution,"* Biology Direct 2, no. 21 (August 20, 2007), https://www.ncbi.nlm.nih.gov/pmc/articles/PMC1973067/pdf/1745-6150-2-21.pdf.

wholesale critique of modern evolutionary theory. He proposes a type of "embryotic evolution," stating, "When a large number of organisms are found all over the earth within a relatively narrow time limit then life had many origins. A large number of different organisms appearing in their final form with all appendages functioning means that all development had occurred before the animals broke ground. That can only mean evolution occurred in a stem cell and that the animal was essentially finished when it appeared in the fossil record. The well-documented inability to discover intermediate forms again points to the stem cell as the evolutionary unit... The literature abounds with stories and glossy pictures of conversion of deer to whale and conversions of fishes to terrestrial animals. While the skeletons depicted are real, the fleshing out and the into—or from— the water migration are indeterminable functions of the Darwinian hypothesis, not of reality. No segment of the record provides evidence of succession in the line of a genotype and there is no evidence for adaptation beyond the limits of the species and the variants that already exist... The fossil record of the evolutionary events supports the conclusions that one must draw at the sight of the Cambrian 'edge' with almost derisive clarity. How much better, Nature would ask, can it be shown that animals do not change appreciably during their existence. Jelly fish are seen in Ediacaran formations dated 600 million years, many arthropods are still alive after 500 million years and modern insects are found frequently as amber inclusions dating back over 100 million years? Are we blindfolded by the paradigm?"[49]

---

[49] Christian Schwabe, *"Embryotic evolution: An ancient question, a new answer,"* Cell Cycle 7, no. 11 (June 1, 2008): 1503– 1506
http://www.tandfonline.com/doi/abs/10.4161/cc.7.11.6357.

CHAPTER SIX

A TEXT CASE STUDY: CAMPBELL BIOLOGY, 11TH EDITION

THE PEARSON WEBSITE CLAIMS THAT THIS TEXT is "the world's most successful majors' biology text."[1] I suspect that this statement is true, and it is probably accurate to say that this text is the most popular choice in colleges and universities for first-year science majors. The authors openly promote a Darwinian view of life. In the section titled "Our Hallmark Features," in the preface of the text, the authors state, "Chief among the themes of both Vision and Change and *Campbell Biology* is evolution. Each chapter of this text includes at least one Evolution section that explicitly focuses on evolutionary aspects of the chapter material, and each chapter ends with an Evolution Connection Question and a Write about a Theme Question."

The first chapter of the book opens with an observation about a mouse's camouflage protection reasoning that the ability of the mouse to hide is, "the result of evolution, the process of change over time that has resulted in the astounding array of organisms found on Earth." The thought continues with, "Evolution is the fundamental principle of biology and the core theme of this book."

Chapter 22 is titled "Descent with Modification: A Darwinian View of Life." In an opening discussion of Aristotle's *scala naturae*, the text summarizes his philosophy, "each form of life, perfect and permanent, had its allotted rung on this ladder. . . these ideas were generally consistent with the Old Testament account of creation, which holds that species were individually designed by God and therefore perfect." A contrast with Darwinian philosophy is then constructed with summarizing statements

---

[1] Pearson. *Campbell Biology 11th edition*. Pearson. [Online] 2017. [Cited: December 31, 2017.] https://www.pearson.com/us/higher-education/program/Urry-Campbell-Biology-Plus-Mastering-Biology-with-Pearson-e-Text-Access-Card-Package-11th-Edition/PGM209351.html.

like, "within a decade, Darwin's book and its proponents had convinced most scientists of the time that life's diversity is the product of evolution."

Chapter 22 continues at this point by describing Darwin's famous "descent with modification." In the section titled as such on page 471, the reader is told that "organisms share many characteristics, leading Darwin to perceive unity in life. He attributed the unity of life to the descent of all organisms from an ancestor that lived in the remote past. . . Darwin reasoned that over a long period of time, descent with modification led to the rich diversity of life we see today."

The reader is next told that "Darwin proposed the mechanism of natural selection to explain the observable patterns of evolution." Then the text gives a concise description of natural selection: "Natural selection is a process in which individuals that have certain heritable traits survive and reproduce at a higher rate than other individuals because of those traits. Over time, natural selection can increase the match between organisms and their environment. If an environment changes, or if individuals move to a new environment, natural selection may result in adaptation to these new conditions, sometimes giving rise to new species."

Clearly the authors of this biology text intend to inform the reader concerning the concepts of Darwinian evolution. One, however, should be careful while reading descriptions of evolutionary biology. The author(s) of this case study text refer to evolution many times as an all-inclusive explanation of the origin and diversity of life. I am not convinced that the Darwinian model sufficiently explains the origin and diversity of life, nor that natural selection has the power to have produced the information necessary to account for the observable life forms in the fossil record. And I am not alone. Nearly 1,000 individuals in the scientific community have publicly, as of January 2016, signed a statement that says, "We are skeptical of claims for the ability of random mutation and natural selection to account for the complexity of life. Careful examination of the evidence for Darwinian theory should be encouraged."[2] Suggesting that the Darwinian model is flawed has harmed the professional careers of many of these scientists.[3]

---

[2] A Scientific Dissent from Darwinism. [Online] 2016. https://dissentfromdarwin.org/.

[3] *Expelled, No Intelligence Allowed.* [DVD] s.l. : Premise Media Corporation, 2008.

This chapter is organized so that each section of the chapter refers to a quotation from the biology text that represents the authors' Darwinian view(s). The fact that most readers will not own the biology text may create challenges. However, descriptions will include quotations from the text and page numbers will be provided. Following each quote from the biology text will be a short discussion about the implications of the statement(s) and some evidence suggesting that the text is indeed written with an intended meaning - a particular view of life. Evidence will be offered in each case that suggests that this Darwinian view of life has flaws that are not discussed.

Example 1: Similarities in Vertebrate Embryos

On page 477 of the text, the caption under an image showing a chick and human embryo (Figure 22.16) says, "At some stage in their embryonic development, all vertebrates have a tail located posterior to the anus (referred to as a post-anal tail), as well as pharyngeal (throat) arches. Descent from a common ancestor can explain such similarities." Darwin used similar arguments.[4]

In the late 1800s, a contemporary of Darwin, Ernst Haeckel, produced drawings of several groups of vertebrate embryos illustrating their similarities and supporting Darwin's idea of common ancestry. The point that Haeckel made with his drawings is that earlier embryos among the different vertebrates are the most similar. In his view, this was evidence that all vertebrates shared a common ancestor. The idea makes sense, however, it was discovered later that Haeckel had fudged his drawings.[5] One famous evolutionary biologist, Stephen Jay Gould, said, "We do, I think, have the right to be both astonished and ashamed by the century of mindless recycling that has led to the persistence of these drawings in a

---

[4] Darwin, Charles. *The Origin of Species by Means of Natural Selection or the Preservation of Favoured Races in teh Struggle for Life.* New York : Bantom Dell a Division of Rnadom House, Inc., 2008 of original text of first edition 1859. pp. 439-440.

[5] *Haeckel, Embryos, and Evolution.* Michael K. Richardson, James Hanken, Lynne Selwood, Glenda M. Wright, Robert J. Richards, Claude Pieau, Albert Raynaud. 5366, s.l. : Science, May 15, 1998, Science, Vol. 280, p. 983.

large number, if not a majority, of modern textbooks.[6] It has been known for more than a century that distinguishing even between chicken and duck embryos as early as day two can be done with ease.[7] Other authors have also challenged the notion that there are early similarities between embryos.[8]

When asked about the incorrect drawings that continue to show up in various forms in both high school and college biology texts, Eugenie Scott, the Director of the National Center for Science Education, said,

> "it's clear that Haeckel may have fudged his drawings somewhat to look more like his ideal than what they actually are. Now does that actually take away from what we know about the relationship of embryology to evolution? Not a bit. The whole Haeckel embryo story has been blown greatly out of significance. It is a minor footnote in the history of science. And actually it has been known for 10 or 15 years that Haeckel's embryos are not to be relied upon. The reason why the diagrams are reproduced is because they are easily available, there's no copyright on them; it's an easy way to illustrate a point. And I would argue that the basic point that's being illustrated by those drawings is still accurate."[9]

Later in the same interview, Dr. Scott still seems to be clinging to the idea that the earlier embryos among vertebrates are the most similar, when it has been known for a long time that the earliest stages are much more dissimilar than the mid stages.[10] Modern texts, of course, don't use

---

[6] Gould, Stephen Jay. *Abscheulich! Atrocious! the Precursor to the Theory of Natural Selection.* Natural History. 2000, Vol. March 2000.

[7] *On the law of development commonly known as von Baer's Law; and on the signficance of ancestral rudiments in embryonic development.* Sedgwick, Adam. s.l. : Quarterly Journal of Microscopical Science, 1894, Vol. 36, pp. 35-52.

[8] Alex T Kalinka, Karolina M Varga et al., *"Gene expression divergence recapitulates the developmental hourglass model,"* Nature 468 (December 9, 2010): 811.

[9] Investigating Evolution A Six Part Educational Series. [DVD] Palmer Lake, Colorado : Coldwater Media, LLC, 2007.

[10] *Haekel's Embryos & Evolution: Setting the Record Straight.* Wells, Jonathan. 5, s.l. : American Biology Teacher, May 1999, Vol. 61, pp. 345-349.

Haeckel's drawings but do select the middle stages of embryological development to make the same point that Haeckel was making.

The author of the *Biology* text has selected an image of a chick and human embryo that look very similar in Figure 22.16 in order to make the case that similarity implies evolutionary relationship. The reader is not told that earlier embryonic images do not look as similar. Then, to exacerbate the situation, a prominent science educator, author and director of the National Center for Science Education says she has no problem with this practice. The intentions of the authors of the biology textbook are clearly to openly promote a Darwinian view of life. The text presentation has omitted information that would be useful for the reader's evaluation of Darwinian philosophy.

Example 2: The Galapagos Finches

Of the examples used in this book, this one is probably the one most people would recognize. So much has written about these birds. The vast majority of biology texts, including the one we are discussing, uses the story of the finches to illustrate the power of natural selection and proof of the validity of a Darwinian view of life.

In the *Campbell Biology* text, an image of the finches is shown in Figure 1.20 on page 16. The text says,

> "Species that are very similar, such as the Galapagos finches, share a common ancestor. Through an ancestor that lived much farther back in time, finches are related to sparrows, hawks, penguins, and all other birds. Furthermore, finches and other birds are related to us through a common ancestor even more ancient. Trace life back far enough, and we reach the early prokaryotes that inhabited Earth over 3.5 billion years ago. We can recognize their vestiges in our own cells – in the universal genetic code, for example. Indeed, all of life is connected through its long evolutionary history."

The author of the text suggests that because the beak sizes change among finches it should be obvious that this mechanism, natural selection, can create all the different species of birds and furthermore all living things on the planet.

Let's take a closer look. Jonathan Weiner published a Pulitzer prize-winning book in the mid-1990s called *The Beak of the Finch: The Story of Evolution in Our Time*.[11] He discusses in detail the changes in beak sizes and speciation that occur among the populations of finches on the Galapagos Islands driven by ecological pressures. Dr. Kenneth Miller, a professor at Brown University, has said that the rate of evolution being witnessed here is fifty to one hundred times faster than Darwin might have expected.[12] However, there are several fundamental questions not being addressed. Aren't they still finches? Is it fair to extrapolate this notion to the evolution of birds in general and then by extension to other vertebrate classes? What happens to the beak sizes when the environmental conditions return to normal?

The actual change in beak sizes over time does not match with the needed observations predicted by the Darwinian model (see figure at top of next page). The beak sizes actually go through cyclical variation based on environmental and other factors and not the expected Darwinian changes.[13]

The answers to the above questions are: Yes, they are still finches. Changing populations of finches does not help to explain the emergence of other vertebrates. The beak sizes return to original forms when the conditions change, and several separate species of finches have been observed interbreeding.[14] It appears to be a cyclical shift in the populations. The extrapolation to essentially explain all vertebrate forms is not supported by this evidence. The changing finch beaks do not offer evidence that the mechanism of natural selection is powerful enough to create the plethora of species on planet earth. The cyclical changes in the

---

[11] Weiner, Jonathan. *The Beak of the Finch: The Story of Evolution in Our Time*. New York : Vintage Books, 1994.

[12] *Investigating Evolution A Six Part Educational Series*. [DVD] Palmer Lake, Colorado : Coldwater Media, LLC, 2007.

[13] Stephen C. Meyer, Scott Minnich, Jonathan Moneymaker, Paul A. Nelson, Ralph Seelke. *Explore Evolution: The Arguments for and Against Neo-Darwinism*. s.l. : Hill House Publishers, 2007. pp. 92-93.

[14] *Unpredictabale evolution in a 30-year study of Darwin's Finches*. Grant, Peter R. Grant and B. Rosemary. s.l. : Science, 2002, Vol. 296, pp. 707-711.

population of finches are not mentioned in the biology text. The authors' intentions to present a Darwinian view of life seem to override the need to present all the evidence.

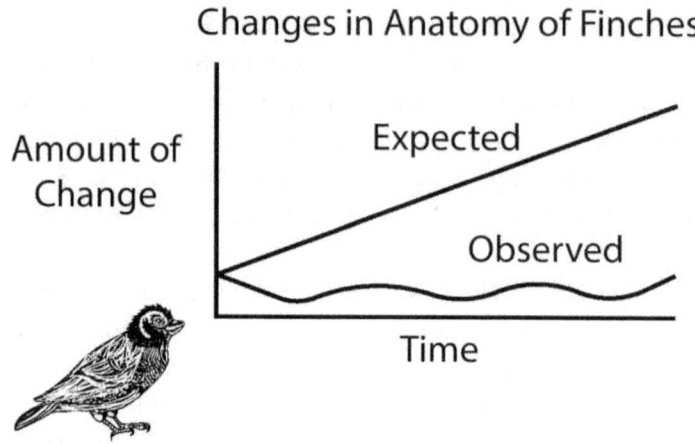

Example 3: Homeotic (*Hox*) Genes and the Development of Animal Bodies

Figure 21.19 on page 461 of the *Campbell Biology* text is titled, "Conservation of homeotic genes in a fruit fly and a mouse." The caption indicates that "Homeotic genes that control the form of anterior and posterior structures of the body occur in the same linear sequence on chromosomes in *Drosophila* and mice." In the text explanation the author says, "Discovering the molecular basis of these differences in turn helps us understand the origins of the myriad diverse forms that cohabit this planet, thus informing our study of the evolution of life." A few sentences later in the text, the author says, "In fact, the nucleotide sequences in humans and fruit flies are so similar that one researcher referred to flies as 'little people with wings.'"

The argument for *Hox* genes from the textbook goes something like this. Gene duplication events could produce new or mutatated *Hox* genes that could influence morphological (like arms, legs, or wings) development. We already know that natural selection acts on random mutation to produce speciation. A major indicator for speciation is change in body plan. So these new or mutated *Hox* genes could result in advantageous changes in body plan, thereby increasing survival rates and reproductive success. So here's a set of genes that could help explain how and why we see different animal body types.

Dr. Kenneth Miller, professor at Brown University, calls these genes part of the tool kit for evolution.[15] He has suggested, and correctly so, that mutations in these genes can cause very rapid and large scale changes in an organism's body. One example that he mentions is a mutation in a *Hox* gene that causes an extra set of wings to grow in fruit flies. He admits that the extra set of wings are not an example of evolution. This admission is a curious statement, given the content of the rest of his statements, which are all about evolution. He says this observation about the fruit flies' wings is an example of the power of the *Hox* gene tool kit.

The *Campbell Biology* text fails to mention that researchers who have successfully made changes in *Hox* genes in Drosophila have shown that these intentional mutations normally cause fatal birth defects.[16] Some researchers have successfully modified *Drosophila* flies with legs growing out of their head[17] or flies with an extra pair of wings.[18] The problem, which should be obvious here, is that these flies are not given an advantage in survival or reproduction. In addition, we also know that *Hox* genes are expressed too late in Drosophila development. About 6,000 cells and the four body axes are already in place before the *Hox* genes turn on.[19]

---

[15] *Investigating Evolution A Six Part Educational Series.* [DVD] Palmer Lake, Colorado : Coldwater Media, LLC, 2007.

[16] *The Molecular Architects of Body Design.* William McGinnis, Michael Kuziora. 2, 1994, Scientific American, Vol. 270, p. 58.

[17] Dan L. Lindsley, E. H. Grell. *Genetic Variations of Drosophila Melanogaster.* s.l. : Carnegie Institute of Washington, 1968.

[18] *Muscle development in the four-winged Drosophila and the role of the Ultrabithorax gene.* J.Fernandes, S.E.Celniker, E.B.Lewis, K. Vijay Raghavan. 11, s.l. : Science Direct, November 1994, Current Biology, Vol. 4, pp. 957-964.

[19]*The segmentation and homeotic gene network in early Drosophila development.* M. P. Scott, S. B. Carroll. 5, Dec 4, 1987, Cell, Vol. 51, pp. 689-98.

If these genes represent evolutionary tool kit, there should be myriads of examples of how *Hox* genes change body plans for the better. In fact, if the *Hox* genes are an example of how evolution works, science labs should be able to produce body plan changes in organisms much more easily. Even if the flies with extra wings could somehow simultaneously grow muscles to make the new pair of wings work, they are still flies. There appears to be some kind of barrier within groups of organisms that will not allow large changes. The Darwinian mechanism does not have an adequate explanation for this barrier.

Once again, the author of the biology text has offered evidence for evolution in a discussion of genes that control the development of animal body plans (*Hox* genes). And once again, the weaknesses in this thinking are not mentioned. The authors' intentions appear to be laser focused on presenting a particular view of life, a Darwinian view.

Example 4: Antibiotic Resistant Bacteria

On pages 476-477, the *Campbell Biology* text says, "An example of ongoing natural selection that dramatically affects humans is the evolution of drug-resistant pathogens (disease-causing organisms and viruses). This is a particular problem with bacteria and viruses because they can produce new generations in a short period of time; as a result, resistant strains of these pathogens can proliferate very quickly." The author goes on to say, "In species that produce new generations in short periods of time, evolution by natural selection can occur rapidly." The point being stressed here is that drug-resistant pathogens are an example of how evolution works.

One of the ways that antibiotics kill bacterial cells is by binding to key molecules (proteins) that are critical for reproduction or sustaining basic life functions of the bacterium. Sometimes a mutation in this binding site can make it impossible for the antibiotic to bind. In that case, an antibiotic resistant species arises. These mutations, however, come at a cost to the bacterial cells. If these mutated cells are asked to grow up competing with regular bacterial cells, they don't survive, something microbiologists call "fitness cost."[20] Sometimes additional mutations help the mutants survive

---

[20] *Is resistance futile?* V.D. Kutilek, D.A. Sheeter, J.H. Elder, B.E. Torbett. 2003, Current Drug Targets - Infectious Disorders, Vol. 4, pp. 295-309.

with greater success, but each new mutation has a cost to the bacterium. The accumulations of mutations never lead to a new type of organism.[21] In fact, British bacteriologist Alan Linton has said, "Throughout 150 years of the science of bacteriology, there is no evidence that one species of bacteria has changed into another."[22]

The Darwinian mechanism, natural selection, does help explain the development of drug resistance in bacteria. However, these resistance observations do not explain the origin of new species or get anywhere close to proving that natural selection acting on random mutation can explain the diversity of living things. Once again, why are these very simple arguments not presented to the reader of the *Campbell Biology* text? The authors' presentation reveals the intention to narrowly focus on only the pro-Darwinian arguments with a glaring absence of any mention of its weaknesses.

Example 5: Anatomical and Molecular Homology

Anatomical: On page 477 of the *Campbell Biology* text, the author says,
> "A second type of evidence for evolution comes from analyzing similarities among different organisms. Characteristics present in an ancestral organism are altered (by natural selection) in its descendants over time as they face different environmental conditions. As a result, related species can have characteristics that have an underlying similarity yet function differently. Similarity resulting from common ancestry is known as homology."

The most common example, and the one used in this text, is an image that shows the upper limbs of a human, a cat, a whale and a bat.

---

[21] *Muller's ratchet decreases fitness of a DNA-based microbe.* Dan I. Anderson, Diarmaid Hughes. January 23, 1996, Proceedings of the National Academy of Sciences, Vol. 93, pp. 906-907.

[22] *Scant search for the maker.* Linton, Alan. April 20, 2001, Times Higher Eduation Supplement, p. 29.

Molecular. On page 478 of the *Campbell Biology* text, the author says, "Biologists also observe similarities among organisms at the molecular level. All forms of life use essentially the same genetic code, suggesting that all species descended from common ancestors that used this code." The text goes on to say that, "humans and bacteria share genes inherited from a very distant common ancestor."

The Darwinian model predicts that homologous structures should be built by homologous developmental pathways and by homologous genes.[23] The body segments of fruit flies and wasps have body segmentation patterns that illustrate homology. These organisms each have an easily recognizable three part segmented body: a head, a thorax, and an abdomen.

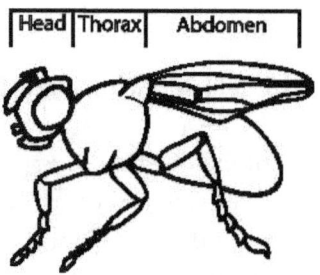

Researchers have shown that the body segments of fruit flies and wasps arise from different developmental pathways.[24] In fact, the body segments of different species of wasps develop along different pathways.[23] Vertebrate limbs, mentioned above, are textbook examples of homologous structures. The development of a variety of vertebrate structures should use homologous pathways and should be directed by homologous genes. However, lamprey guts develop from the floor of the embryonic cavity, while shark guts develop from the roof of the embryonic cavity and frog

---

[23] Alfred S. Romer, Thomas S. Parsons. *The Vertebrate Body*. Philadelphia : W.D. Saunders, 1977. pp. 9-10.

[24] *Evo-Devo aspects of classical and molecular data ina historical perspective*. Klaus Sander, Urs Schmidt-Ott. 2004, Journal of Experimental Zoology B (Molecular and Developmental Evolution), Vol. 302, pp. 69-91.

guts from both the roof and the floor of the embryonic cavity.[25] These observations do not match the Darwinian predications of homology.

Researchers have also shown that similar genes help to produce very different types of structures in some species.[26] The *pax6* (called *ey* in flies) gene is critical in building the eyes of mice, squid, and fruit flies. The mouse and squid eye both have one lens, but develop along different pathways and are connected to the nervous system differently. The eye of the fruit fly has dozens of lenses. This illustration points out that the same gene helps in the development of all three of these eyes, a very non-Darwinian expectation.

In the next section of the *Campbell Biology* text, page 478, the author says, "Some homologous characteristics, such as the genetic code, are shared by all species because they date to the deep ancestral past. In contrast, homologous characteristics that evolved more recently are shared only within smaller groups of organisms." The text goes on then to explain how homology helps in constructing evolutionary trees. These trees "represent the pattern of descent from common ancestors." These representations of evolutionary relationships among living things are common in general biology textbooks. They are often called phylogenetic trees. Phylogenetics is a process of identifying evolutionary relationships. Some of the most popular techniques in modern research are to compare DNA sequences - or anatomical characteristics (or both) - to uncover common ancestry.

The reader of the *Campbell Biology* text is not told that evolutionary trees constructed using biological molecules like DNA or RNA often don't resemble those made using anatomical structures.[27] The reader is not told

---

[25] Stephen C. Meyer, Scott Minnich, Jonathan Moneymaker, Paul A. Nelson, Ralph Seelke. *Explore Evolution: The Arguments for and Against Neo-Darwinism.* s.l. : Hill House Publishers, 2007. pp. 44.

[26] *Developmental functions of the Distal-less/Dlx homeobox genes.* Grace Panganiban, John L.R .Rubenstein. 2002, Development, Vol. 129, pp. 4371-4386.

[27] A) *Bones, Molecules or Both?* Gura, T. July 2000, Nature, Vol. 406, pp. 230-233. The complete mitochondrial DNA sequence of shark Mustelus manazo: Evaluating rooting contradictions to bony vertebrates. Ying Cao, Peter J. Waddell, Norhiro

that conflict in the production of these trees is the norm and not the exception, which has led some scientists to conclude that the evolutionary tree concept is obsolete and should be discarded.[28] The reader is additionally not told about the groundbreaking research in microRNAs leading Dartmouth biologist Kevin Peterson to say, "I've looked at thousands of microRNA genes, and I can't find a single example that would support the traditional tree."[29]

Presumably, only one real evolutionary tree can be correct. These conflicting results from researchers pose a serious problem to the Darwinian model, and none of this is mentioned to the reader of the *Biology* text.

The author's intentions in the *Campbell Biology* text are once again obvious. Every effort is being made to point the reader toward a particular

---

Okado, Masami hasegawa. 1998, Molecular Biology and Evolution, Vol. 15, pp. 1637-1646.

B) *18S gene trees are positively misleading for monocot/dicot phylogenetics*. Melvin R. Duvall, Autumn Bricker Ervin. 2004, Molecular Phylogenetics and Evolution, Vol. 30, pp. 97-106.

C) *Molecules vs. morphology in avian evolution: The case of the 'pelecaniform' birds*. S. Blair Hedges, Charles G. Sibley. October 1994, Proceedings of the National Academy of Sciences USA, Vol. 91, pp. 9861-9865.

D) *Conflicting phylogenetic signals at the base of the metazoan tree*. A. Rokas, N. King, J. Finnerty, S.B. Carroll. 2003, Evolution and Development, Vol. 5, pp. 346-359.

[28] A) *Understanding Phylogenetic Incongruence: Lessons from Phyllostomid Bats*. L.M. Dávalos, A.L. Cirranello, J.H. Geisler, N.B. Simmons. 4, 2012, Biological Reviews of the Cambridge Philosophical Society, Vol. 87, pp. 991-1024.

B) *Why Darwin Was Wrong about the Tree of Life*. Lawton, G. January 21, 2009, New Scientist, pp. 34-39.

C) *Phylogenetic Classification and the Universal Tree*. Doolittle, W. F. June 25, 1999, Science, Vol. 284, pp. 2124-2128.

[29] *Rewriting Evolution*. Dolgin, E. June 28, 2012, Nature, Vol. 486, pp. 460-462.

way of seeing the world. The absence of easily accessible data from the scientific literature that undermines the Darwinian model is glaringly absent.

Example 6: The Cambrian Explosion

On page 534 of the *Campbell Biology* text, the author says, "Many present-day animal phyla appear suddenly in fossils formed 535-525 million years ago early in the Cambrian period. This phenomenon is referred to as the Cambrian explosion." The text goes on to say, "In a relatively short period of time, predators over 1 meter in length emerged that had claws and other features for capturing prey; simultaneously new defensive adaptations such as sharp spines and heavy body armor, appeared in their prey." Although the text does not say, the author apparently recognizes the obvious problem that these observations pose to the Darwinian model. The next paragraph mentions that the extant mollusc, *Kimberella*, dates to 560 million years and that some DNA analyses indicate living sponges at 700 million years ago. These observations do very little to alleviate the glaring problem of the sudden appearance in the fossil record of "many present-day animal phyla."

In Darwin's book, chapter 9 is titled "On the Imperfection of the Geological Record." In the subsection titled, "On the Sudden Appearance of Whole Groups of Allied Species," he says,

> "The abrupt manner in which whole groups of species suddenly appear in certain formations, has been urged by several paleontologists, for instance, by Agassiz, Pictet, and by none more forcibly than by Professor Sedgwick, as a fatal objection to the belief in the transmutation of species. If numerous species, belonging to the same genera or families, have really started into life all at once, the fact would be fatal to the theory of descent with slow modification through natural selection. For the development of a group of forms, all of which descended from some one progenitor, must have been an extremely slow process; and the progenitors must have lived long ages before their modified descendants."[30]

---

[30] Darwin, Charles. *The Origin of Species by Means of Natural Selection or the Preservation of Favoured Races in the Struggle for Life*. New York : Bantam Dell a Division of Random House, Inc., 2008 (1859). pp. 296-297.

Paleontologists indicate that the Cambrian lasted about 10 million years, and that this extent of time is not enough for the Darwinian mechanisms to produce what has been found in the fossil beds.[31] On page 531 of the *Campbell Biology* text, an image of the geological record shows the sudden appearance of mammals during the Paleocene epoch (65-56 million years ago). On this same image, flowering plants are shown suddenly emerging during the Cretaceous period, which has been pointed out as a major problem for the Darwinian model since they have "no obvious ancestors for a period of 80-90 million years before their appearance."[32]

The *Campbell Biology* text, on page 534, says, "Prior to the Cambrian explosion, all large animals were soft-bodied." Although not stated directly, this observation appears to imply that since they were soft-bodied, we would not expect to find them in the fossil beds. This expected lack of discovery does not represent an adequate explanation, as recent research in Cambrian strata has produced fossilized soft-bodied animals (including soft-bodied embryos) belonging to several different animal groups.[33] If

---

[31] A) *Calibrating rates of early Cambrian evolution.* S.A. Bowring, J.P. Grotzinger, C.E. Isachsen, A.H. Knoll, S.M. Pelechaty, P. Lolosov. 1993, Science, Vol. 261, pp. 1293-1298.

B) *A new look at evolutionary rates in deep time: Uniting paleontology and high-precision geochonology.* S.A. Bowring, J.P. Grotzinger, C.E. Isachsen, A.H. Knoll, S.M. Pelechaty, P. Lolosov. 1998, GSA Today, Vol. 8, pp. 1-8.

C) *Bayesian models of episodic evolution support a late Precambrian explosive diversification of the Metazoa.* S. Aris-Brosou, Z. Yang. 2003, Molecular Biology and Evolution, Vol. 20, pp. 1947-1954.

D) *Towards a new evolutionary synthesis.* Carroll, Robert L. January 2007, Trends in Ecology and Evolution, Vol. 15, pp. 27-32.

E) *The notion of the Cambrian panamalia genome.* Ohno, Susumo. August 1996, Proceedings of the National Academy of Sciences, Vol. 93, pp. 8475-8478.

[32] *Genome duplication and the origin of angiosperms.* Stefanie De Bodt, Steven Maere, Yves Van de Peer. 2005, Trends in Ecology and Evolution, Vol. 20, pp. 591-597.

there were soft-bodied ancestors to the Cambrian animals, they would have been found.

The *Campbell Biology* text does not mention these significant fossil findings, suggesting once again the intent to present only the Darwinian model. In my opinion, the authors of this science text intentionally omit these observations in order to prevent the readers from questioning the Darwinian model.

Example 7: DNA and the Genetic Code

On pages 341-342 of the *Campbell Biology* text, the author says, "The genetic code is nearly universal, shared by organisms from the simplest bacteria to the most complex plants and animals." A couple of examples are given for illustration purposes, and then the text says, "Despite a small number of exceptions, the evolutionary significance of the code's near universality is clear. A language shared by all living things must have been operating very early in the history of life – early enough to be present in the common ancestor of all present-day organisms."

By "near" universality, I'm assuming that the author is referring to easily accessible information available through the National Center for Biotechnology Information, which currently lists 24 different genetic codes

---

[33] A) Valentine, James W. *The Macroevolution of Phyla; Soft-bodied Fossils.* [ed.] Phillip W. Signor Jere H. Lipps. *Origin and Early Evolution of the Metazoa.* New York : Plenum Press, 1992, pp. 525-553; 529-531.

B) Paul Chien, J.Y. Chen, C.W. Li, Frederick Leung. *SEM Observation of Precambrian Sponge Embryos from Southern China, Revealing Ultrastructures Including Yold Granules, Secretion Granules, Cytoskeleton and Nuclei.* s.l. : North American Paleontological Convention University of California, Berkeley, 2001.

C) *Cellular and subcellular structure of neoproterozoic animal embryos.* J.W. Hagadorn, S. Xiao, C. J. Donoghue, S. Bengtson, N.J. Gostling, M. Pawlowska, E.C. Raff, R.A. Raff5, F.R.Turner, Y. Chongyu, C. Zhou, X. Yuan, M.B. McFeely, K.H. Nealson, M. Stampanoni. October 2006, Science, Vol. 314, pp. 291-294.

in a range organisms from yeasts to vertebrates.[34] Many of these vary significantly from the "universal" code.[35] The authors of the book, "Explore Evolution," summarize the issue as follows: "Once a code is established, it can't change without destroying the organisms that depend on it. Any small change in the codon-amino acid assignments means that the right amino acids will not be added in the right place. The resulting chains might not fold properly, and might not function as a protein. This could be disastrous for the organism."[36] This codon language will make more sense as you keep reading, but for now simply notice that the observation of a variety of codes among living things does not fit a Darwinian view of life. Also note that none of this information is made available to the reader of the *Campbell Biology* text.

In order to fully appreciate what is meant by the Biology text's statement "a language shared by all living things," one will need some basic understanding of DNA and cellular mechanisms that lead to the formation of proteins.

Human beings are said to be carbon based life forms, and this notion is true. The human body is made of cells, over 300 trillion of them. And these cells are constructed of molecules with carbon as the central core of the structure. Each of these cells, with the exception of mature red blood cells, has a control center in it called a nucleus, arguably the most important organelle. Other organelles include things like mitochondria (which also contain DNA) for making energy and endoplasmic reticulum for processing proteins. The nucleus is the brain of the cell. Within the nucleus of human cells is the carbon-based molecule called DNA. The fancy science word is deoxyribonucleic acid.

---

[34] Information, National Center for Biotechnology. The Genetic Codes. [Online] November 2016.
https://www.ncbi.nlm.nih.gov/Taxonomy/Utils/wprintgc.cgi?mode=c.

[35] *How unique is the genetic code.* Christine Fenske, Gottfried J. Palm, Winfried Hinrichs. 2003, Angewandte Chemie International Edition, Vol. 42, pp. 606-610.

[36] Stephen C. Meyer, Scott Minnich, Jonathan Moneymaker, Paul A. Nelson, Ralph Seelke. *Explore Evolution: The Arguments for and Against Neo-Darwinism.* s.l. : Hill House Publishers, 2007. pp. 54.

Compared to the structure of proteins, DNA is pretty simple to describe. It's kind of like a ladder that has been twisted. The sides of the ladder are repeating units of a five carbon sugar, called deoxyribose, and phosphate groups. So, as you go down the sides of the twisted ladder, you find sugar-phosphate-sugar-phosphate-etc. The rungs of the ladder are built out of nitrogen-containing bases. Four of them are used to build DNA. They are adenine (A), cytosine (C), guanine (G), and thymine (T). Each rung will either be an adenine paired to a thymine or a cytosine paired to a guanine. These are called base pairs. Each base pairs with its complementary base by way of weak hydrogen bonds. These bonds are horizontal but not vertical. Each base hangs onto the sides of the ladder by way of a chemical bond (called an N-glycosidic bond). Now, even though DNA structure is pretty simple, it has lots and lots of letters, about three billion. Just to give you some perspective on how many letters this is, if you counted out loud at a normal rate to three billion, it would take you about 300 years.

DNA

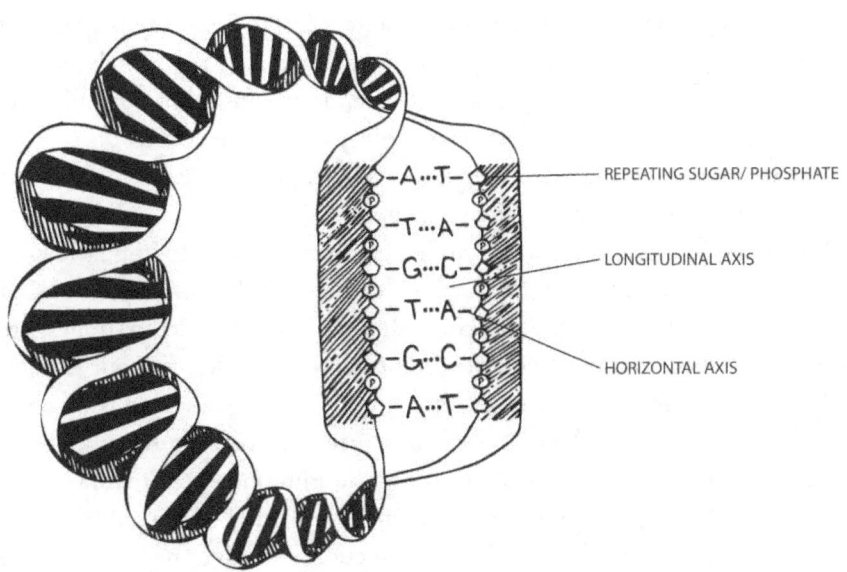

So, now you know the basics of the structure of DNA. So what is meant by saying that DNA functions as the brains of the cell? In science, the connection of DNA to protein is referred to as the central dogma. The central dogma refers to the fact that part of the DNA is copied into RNA (ribonucleic acid), which then facilitates the formation of proteins. Proteins are very complicated molecules that are built using amino acids, like methionine, valine, etc.

Making Proteins: DNA sequence: TAC   CAA
RNA sequence: AUG   GUU
Amino acids: Met--------Val (growing protein)

There are twenty amino acids that are used to build proteins in humans. Look at the three step summary above and find where the amino acid methionine (Met) joins to valine (Val) as the first two amino acids in this growing protein. The amino acids are being put together using instructions from the DNA. The instructions are found in three base sequences, or three letter sequences. So, the three letter sequence TAC in the DNA sequence is a code for AUG in the RNA sequence and finally, the amino acid methionine (Met). CAA then tells valine to be added next, and on and on the story goes. If one of the letters in the sequence is wrong, it can cause major problems in the proper construction of the protein/amino acid sequence. For example, sickle-cell anemia is caused by a problem in the building of a protein called hemoglobin (oxygen transporter). A single base (letter) substitution, also sometimes called a point mutation, causes the wrong amino acid to be put in the protein.

Proteins are really important. It is very difficult, next to impossible, to talk about any structure or function in the human body without talking about proteins. For example, we know that the human skeleton, bones, are nearly one-third by weight collagen, a protein. The flesh of our bodies, skeletal muscle, and our hearts (cardiac muscle) are by weight mostly actin and myosin, proteins. Physiologists must use the language of proteins in describing the how the brain works, how the immune system works, how kidneys make urine, how muscles contract, and on and on we could go.

Genes are normally described as sections (coding regions) of the DNA/RNA molecules that give instructions for how to build proteins. This idea is what we mean when we say central dogma. The genetic code is just that - a language of three-letter (bases) codes that contain the instructions on how to build proteins.

Clearly, the three letter sequences in the DNA molecule that ultimately dictate which amino acid gets added to a growing protein are a language that the cells understand, a language critical for life. So, explaining how the letters end up in their particular order is critical in understanding both the origin and diversity of life. If there are chemical forces (natural causes) that can explain the order of the letters, then we are much closer to explaining life in general.

Stephen Meyer, in his book "Signature in the Cell," explains the significance of the order of the DNA base letters in chapter 11, titled "Self-Organization and Biochemical Predestination."[37] On page 243, he says, "There are also hydrogen bonds stretching horizontally across the molecule between nucleotide bases, forming complementary pairs. . . But notice too that there are no chemical bonds between the bases along the longitudinal axis in the center of the helix."

He is pointing out that there is no chemical bonding (no natural explanation) that dictates in the language that one particular letter follows another along the longitudinal axis of the molecule. Meyer goes on to say, "There are no significant differential affinities between any of the four bases and the binding sites along the sugar-phosphate backbone. Instead, the same type of chemical bone (an N-glycosidic bond) occurs between the base and the backbone regardless of which base attaches. All four bases are acceptable, none is chemically favored." In other words, there is no chemical bonding explanation for the order of letters either in the longitudinal or the horizontal axes of the DNA molecule. The language of the genetic code was clearly written by an intelligence. This fact has driven some scientists to suggest that life arrived on the planet already functioning and "evolved" somewhere else.[38] The implication is that aliens seeded life here. The idea of aliens cannot be considered serious science. This problem

---

[37] Meyer, Stephen. *Signature in the Cell, DNA and the Evidence for Intelligent Design*. New York : HarperCollins Publishers, 2009.

[38] Wickramasinghe, Chandra and Bauval, Robert, *Cosmic Womb, the Seeding of Planet Earth* Theory, Rochester, Vermont: Bear and Company Publishers, 2017

with the Darwinian model has not escaped the notice of many prominent scientists,[39] but is noticeably omitted from the *Campbell Biology* text.

Example 8: Evolutionary Origins of Bacterial Flagella

Flagella are structures that enable many types of bacterial cells to swim, acting somewhat like a rotary propeller. On pages 574-575 of the *Campbell Biology* text, the author says,

> "The bacterial flagellum shown in Figure 27.7 has three main parts (the motor, hook and filament) that are themselves composed of 42 different kinds of proteins. How could such a complex structure evolve? In fact, much evidence indicates that bacterial flagella originated as simpler structures that were modified in a stepwise fashion over time. . . Analyses of hundreds of bacterial genomes indicate that only half of the flagellum's protein components appear to be necessary for it to function . . . Of the 21 proteins required by all species studied to date, 19 are modified versions of proteins that perform other tasks in bacteria. For example, a set of 10 proteins in the motor is homologous to 10 similar proteins in a secretory system found in bacteria. . . The proteins that comprise the rod, hook, and filament are all related to each other and are descended from an ancestral protein that formed a pilus-like tube. These findings suggest that the bacterial flagellum evolved as other proteins were added to the ancestral secretory system. This is an example of exaptation, the process in which structures originally adapted for one function take on new functions through descent with modification."

This section on the evolution of the bacterial flagella follows a discussion of motility where the author says,

> "Overall, these structural and molecular comparisons indicate that the flagella of bacteria, archaea, and eukaryotes arose independently. Since current evidence shows that flagella of organisms in the three domains perform similar functions but are

---

[39] A Scientific Dissent from Darwinism. [Online] 2016. https://dissentfromdarwin.org/.

not related by common descent, they are described as analogous, not homologous, structures."

The textbook goes on to state, "Only half of the flagellum's protein components appear to be necessary for it to function." So, twenty one of the proteins are required. Research, however, highlights the fact that if any one of these proteins is "knocked out," the bacterial cell loses the ability to swim.[40] This means that these twenty one proteins need to all be in their proper place at one time for the flagella to function. The explanation offered in the text does not address this problem. Instead, the author suggests to the reader that the Darwinian mechanism can explain this because nineteen of the twenty one required proteins perform other tasks in the bacteria. We, the readers, are asked to make the assumption that these proteins somehow mutated into the ones now found in the flagella. An example is then offered. Of the forty two proteins in the flagella, ten of them are very similar to ten of the proteins in a secretory system in bacteria. So, the reader is to assume that proteins of the secretory system became proteins of the flagella in a Darwinian fashion.

Bacteria are prokaryotes and use their secretory system to inject poison into eukaryotic cells in a pathogenic way. This event disrupts cellular processes in the injected cell.[41] Research has shown that the flagella is actually older than the secretory system, which if true, makes the exaptation argument seem silly.[42] The *Campbell Biology* text says that prokaryotes emerge in the Archaean eon long before the eukaryotes do during the Proterozoic eon. In an evolutionary argument, it makes more sense that the ability to move (having a flagella) would have evolved before a secretory system (used to inject Eukaryotes, which weren't around for millions of

---

[40] *Diverse high-torque bacterial flagellar motors assemble wider stator rings using a conserved protein scaffold.* Morgan Beeby, Deborah A. Ribardo, Caitlin A. Brennan, Edward G. Ruby, Grant J. Jensend, David R. Hendrixson. 13, 2016, Proceedings of the National Academy of Sciences, Vol. 113, pp. E1917-E1926.

[41] *Type III Protein Secretion Systems in Bacterial Pathogens of Animals and Plants.* Hueck, Christoph J. 2, June 1998, Microbiology and Molecular Biology Reviews, Vol. 62, pp. 379-433.

[42] *The Non-Flagellar Type III Secretion System Evolved from the Bacterial Flagellum and Diversified into Host-Cell Adapted Systems.* Sophie S. Abby, Eduardo P. C. Rocha. September 27, 2012, PLOS: Genetics.

years). Also, mutational density studies, phylogenetic distribution analyses, and plasmid analyses observations all agree with each other that the flagella are evolutionarily older than the secretory systems.[43]

As it turns out, a number of protein tools are also required in the construction of the bacterial flagella. If any one of these tools is "knocked out," then the flagella will not be made properly.[44] In short, the flagella does not work without all its parts in place, and the system that makes the flagella does not work without all its parts in place.

Michael Behe summarizes the problems posed here to the Darwinian model by saying,

> "In summary, as biochemists have begun to examine apparently simple structures like cilia and flagella, they have discovered staggering complexity, with dozens or even hundreds of precisely tailored parts. As the number of parts increases, the difficulty of gradually putting the system together skyrockets . . . New research on the roles of auxiliary proteins cannot simplify the irreducibly complex system. The intransigence of the problem cannot be alleviated; it will only get worse."[45]

Once again, none of this information is made available to the reader of the *Campbell Biology* text, revealing an intention to teach the readers a particular Darwinian view of life.

---

[43]Klinghoffer, David. *Refuting Behe's Critics, Meyer Gives Four Reasons the Flagellum Predates the Type III Secretory System.* Evolution News. [Online] November 2016. https://evolutionnews.org/2016/11/stephen_meyer_g/.

[44]Scott A. Minnich, Stephen Meyer. *Genetic analysis of coordinate flagellar and type III regulatory circuites in pathogenic bacteria.* [book auth.] Stephen C. Meyer Scott A. Minnich. [ed.] M.W. Collins. *Design and Nature II: Comparing Design in Nature with Science and Engineering.* s.l. : Southampton: WITpress, 2004, pp. 295-304.

[45] Behe, Michael J. *Darwin's Black Box.* New York : Touchstone by Simon & Schuster, 1996. p. 73.

ABOUT THE COVER

In order to build new anatomical structures (e.g., body plans) or any new functional system (e.g., reproductive, digestive), an enormous amount of new genetic information is required. Much of this information is in the DNA molecule. It is stored in three base sequences that we call the genetic code.

Many different lines of reasoning have been used to explain the origin of the genetic code for constructing new animal forms in the context of the origin of life. They include discussions of mutation, gene duplication, recombination, the RNA world hypothesis, pseudogenes, *Hox* genes, junk DNA and others. As this problem has become more widely recognized, new explanations are emerging. They include things like evolutionary development biology, self-organization, natural genetic engineering, neutral evolution and neo-Lamarckism/epigenetic inheritance.

What appears to be obvious to many scientists today is that the simplistic model of life envisioned by the Darwinian modern synthesis does not get anywhere near explaining the origin of the sequence of bases in a single DNA molecule. And it does not offer, in my opinion, a satisfactory explanation for where all the new information comes from in each group of living things.

The cover contains an image of the DNA molecule highlighting the sequence of letters that could be part of a gene. The rest of the cover contains many of the three base sequences of DNA that code for particular amino acids (written beside their codes) in the construction of proteins.

www.ingramcontent.com/pod-product-compliance
Lightning Source LLC
Chambersburg PA
CBHW070203230526
45471CB00002B/801